DATE DUE

NO 17 98		
DE 8'08		
SE 27 99		

DEMCO 38-296

YOU ARE A
MATHEMATICIAN

R

YOU ARE A MATHEMATICIAN

A Wise and Witty Introduction to the Joy of Numbers

David Wells

John Wiley & Sons, Inc.
New York • Chichester • Weinheim • Brisbane • Singapore • Toronto

This text is printed on acid-free paper.

Copyright © 1995 by David Wells
Published by John Wiley & Sons, Inc.
All rights reserved. Published simultaneously in Canada.

First published in 1995 by Penguin Books U.K.

This publication is designed to provide accurate and authoritative
information in regard to the subject matter covered. It is sold with the
understanding that the publisher is not engaged in rendering legal,
accounting, or other professional services. If legal advice or other expert
assistance is required, the services of a competent professional person should
be sought.

Library of Congress Cataloging-in-Publication Data
Wells, D. G. (David G.)
    You are a mathematician : a wise and witty introduction to the joy
of numbers / David Wells.
        p.   cm.
    Includes index.
    ISBN 0-471-18077-7 (cloth : alk. paper)
    1. Mathematics--Popular works. I. Title.
    QA93.W385 1997
    510--dc21   96-53548

Printed in the United States of America

10 9 8 7 6 5 4 3 2 1

Contents

Acknowledgements

To produce a book of this complexity requires the enthusiastic, even devoted, assistance of many people. I should like to thank John Sharp who was once again responsible for organizing the many diagrams, and drawing most of them, and Rodney Paull who drew the remainder; John Bentin for his meticulous copy-editing and attention to detail, which improved the text in many places; and my editor, Ravi Mirchandani, together with Georgina Widdrington and Andrew Cameron and the staff at Penguin Books. Whatever errors and infelicities remain are my sole responsibility.

Finally, I am as always extremely grateful to David Singmaster for placing the resources of his library at my disposal.

Introduction

'Steve Davis is walking round the table. He has a choice of pots, none of them very difficult. He's decided to go for the red by the middle pocket and screw back for the pink . . . Oh dear, that was a bad mistake, he's left it open for his opponent with all the reds in good positions . . .'

As Steve Davis makes his mistake a million viewers gasp, then wait with knowledgeable anticipation for his opponent to take advantage. Why is snooker so easy to appreciate? Why is it ideally suited for television? Why do so many people enjoy watching the game? Snooker is extraordinarily visual. You can see so much of what is happening. It's all above board. The only thing you cannot usually see is precisely how the players strike the cue ball with the tip of the cue, the very subtlety that allows them to control the cue ball – but you don't have to see this in order to appreciate the game, especially when a commentator is at hand.

Mathematics is the complete opposite of snooker. Watch a world-class mathematician doing mathematics and you could see nothing at all, except someone sitting at a desk thinking, or pacing up and down a room, or going for a walk, or reading a book. Even if they were writing or drawing on a blackboard, you probably wouldn't have any idea what they were writing or why, or what it meant.

Mathematics, as teachers are fond of repeating, is *not a spectator sport*. It's exciting, it's fascinating, it's mysterious, but you can't just put it up in vivid colours on a TV screen and expect viewers to grasp it while sitting back in an easy chair. At least, the reader of a mathematics book must be prepared to think about what the author is saying. At best the reader gets stuck in and becomes involved with the author's ideas. Professional mathematicians do this all the time. When they read a book, or a paper by a colleague, they reach for a pen and paper and they start to work out the ideas for themselves, because they know that 'To do is to understand'.

This book gives you, the reader, the chance to do just that. Each chapter is interrupted by a number of boxes which pose questions to you, related to the theme of the chapter. Some are easy, some are more difficult, but they will all make you think. Tackle them, and you too will be a Mathematician. Of course, if you don't feel like exercising your grey cells, you can always look up the solutions, but I hope that you will feel drawn to the problems, and experience an irresistible temptation to attempt to solve them! Good luck, and *bon voyage* on your mathematical journey!

Chapter 1
The hidden world of triangles

Problem 1A   *Across the river*

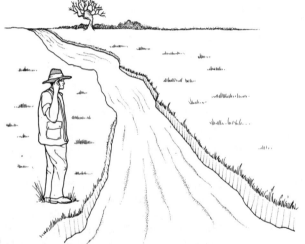

Fig. 1.1

Jake Hardy was standing on the river bank, looking across to the far side. 'How wide do you reckon it is?' asked Harold. Jake adjusted the rim of his hat, and turned to look downstream. He paused and then walked with deliberate paces along the river bank, then turned and called out, 'About thirty metres, give or take a few.'

How did he estimate the width of the river? (Solution on page 22)

Everyone is familiar with the simplest mathematical objects, such as straight lines and circles and squares, and the counting numbers. To be a mathematician all you have to do is to learn to look at these objects with some insight and imagination, maybe do a few experiments too, and be able to draw reasonable conclusions.

Fig. 1.2

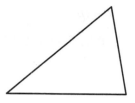

The result of these activities – which are also quite familiar to you from everyday life – is that you soon see the square as more than something with four equal sides and four right angles; a circle as much more than, well, just a plain circle; and the number 8 as much more than merely the next number after 7. We are going to start this mathematical adventure by looking at another very simple and common mathematical object, the humble triangle.

Figure 1.2 shows a naked triangle in a vast empty plane which you must imagine extends for ever, far beyond the edges of this page. What can we say about it? What features does it have? The three sides could be any length at all – except that the two shorter sides together must be longer than the longest side, or the triangle would not close. You cannot make a triangle out of three strips of wood of length 3, 5 and 12 metres.

With that proviso, any three lengths will make a unique rigid triangle. It can be turned over, or turned round, but it is essentially 'the same'. This fact is useful to surveyors, because, if you know the three sides of a triangle formed by three landmarks, then you can draw a map which shows the exact relative positions of the land-marks, but on a smaller scale. By repeated *triangulation*, a large area can be mapped accurately from a small number of landmarks. It is also useful to engineers, who make the triangle the basis of their structures. There is no need to weld the joints on the frame shown in Fig. 1.3; it will be rigid if the joints are merely pinned.

The three angles cannot be chosen as freely as the three sides. In fact, when we know the size of two of them, the other one can be

Fig. 1.3

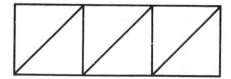

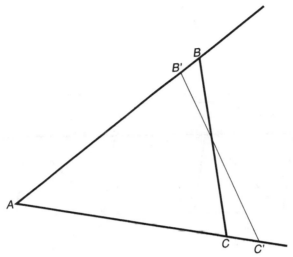

Fig. 1.4

calculated, because their sum is constant. This might be suggested by this small experiment. Imagine that two sides of a triangle are fixed, and the third is rotating slowly, as indicated by the thin line in Fig. 1.4.

Problem 1B   *Bars and joints*

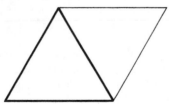

Fig. 1.5

A plane frame made of bars and pinned joints which is just rigid can be constructed by starting with a bar, and then adding more bars, two at a time, so that each extra pair creates a triangle (Fig. 1.5).

What is the relationship between the number of bars and the number of joints?

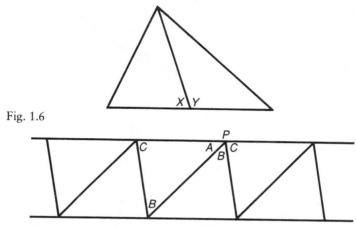

Fig. 1.6

Fig. 1.7

It seems plausible that, as BC rotates to a new position $B'C'$, then the angle between BC and AB has increased by just the same amount as the angle between BC and CA has decreased, keeping their sum constant. If we suspect that their sum is constant, we can see what it must be from Fig. 1.6. There are three triangles in the figure, and the angles of the larger triangle are the sum of all the angles of the two smaller triangles, excluding X and Y. If the angle sums of all three triangles are equal, then the angle sum of each of the smaller triangles must be $X + Y$. This is also suggested by the frieze shown in Fig. 1.7, which is simply a strip of paper divided by successive parallel lines, alternately in two directions, into apparently identical triangles. At a point such as P, we find one of each of the angles of the triangle represented, so it seems once again that $A + B + C$ is constant: that is, $A + B + C = 180°$.

These simple physical experiments – so simple that we might even be able to do them in our heads as 'thought experiments' – may seem convincing, but they have a hidden weakness which points to the need for further investigation.

In Fig. 1.8, a 'triangle' has been drawn on a sphere, as if B were the north pole and ACC' the equator. In this case, as the side BC rotates to a new position BC', the angle between BC and AB increases, but the angle between BC and CA remains the same!

So it seems that, on a sphere, the angle sum of a triangle can be increased by rotating one of the sides. Somewhere, somehow, there is

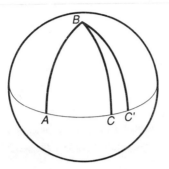

Fig. 1.8

a difference between the plane paper and the curved surface of the sphere which explains why angle sums of triangles are constant on one, but not on the other.

✳

Problem 1C   *Polygon angles*

How does Fig. 1.9 suggest that, *if* the angle sum of any quadrilateral is a fixed quantity, it must be 360°?

Fig. 1.9

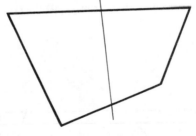

How does Fig. 1.10 suggest that, *if* the angle sum of a polygon is fixed by the number of sides, then adding a side increases the angle sum by 180°?

Fig. 1.10

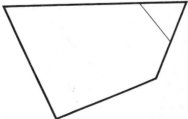

What other properties does the original triangle have? None, until we use our imaginations and start asking some searching questions. As soon as we start to pose problems, and to solve them, we inevitably find ourselves discovering more of its many features. One natural question is: how big is this triangle? What is its area? The simplest and traditional way to find the area is to divide the triangle into two right-angled triangles, by drawing an *altitude*, as in Fig. 1.11, and then drawing a horizontal line which bisects the altitude at right angles. This dissects each right-angled triangle into a rectangle. In a total of three pieces, the original triangle has been transformed into a rectangle of the same height, and half the width. This trick can be performed in three different ways, if all three angles of the triangle are less than a right angle, by starting with each of the three sides (Fig. 1.12). This at once tells us something about the lengths of the altitudes and the sides:

$$BC \times AD = CA \times BE = AB \times CF.$$

Is there any other connection between the three altitudes? The mathematician will naturally suspect that there may be, and a small test is worth carrying out. Sure enough, they appear to *concur* (meaning that they pass through a common point). Indeed, they do, in a point called the *orthocentre* of A, B and C (see Fig. 1.13).

Fig. 1.11

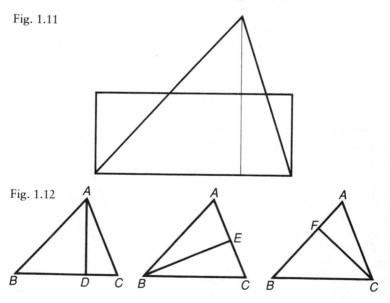

Fig. 1.12

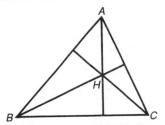

Fig. 1.13

There is a further small bonus awaiting us here, provided that we keep our eyes open. Suppose that we had started with the triangle *HBC*. The three altitudes would then have been *AB*, *AC* and *AH*. They concur, of course, at *A*. In other words, the relationship between *A, B, C* and *H* is symmetrical. Each point of the four is the orthocentre of the other three. What an elegant and simple relationship!

By asking a simple question about areas, and by keeping our eyes peeled, we have been naturally led to discover some new facts about triangles. To the mathematician, this is expected and quite natural. It is actually very difficult in mathematics to ask novel questions which do *not* lead to new discoveries.

*

If the triangle were a real physical sheet, made of some uniform material, it would not only have an area, but also a *centre of gravity*; see Fig. 1.14. This is the point on which the triangular sheet would balance on a pinpoint. If the triangle were suspended from a vertex, a vertical line through that vertex will also pass through the centre of gravity. If the triangular sheet were resting on the edge of a table, it would start to tip and fall over the edge, if its centre of gravity were over the edge. Where is the centre of gravity? If the triangle is divided into numerous narrow parallel strips, each strip will balance about its midpoint, and all these midpoints appear to lie on the straight line

Fig. 1.14

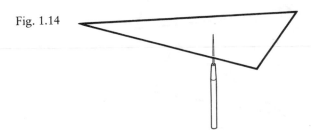

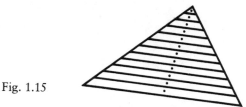

Fig. 1.15

joining the vertex to the midpoint of the opposite side (Fig. 1.15), called the *median* from that vertex. The centre of gravity of all the strips together will lie somewhere on the same straight line. Bearing in mind that we found the area of the triangle in three different ways by starting with each side as base in turn, it is natural to do the same for the centre of gravity of the triangle. Figure 1.16 shows the three constructions, each with a line of midpoints. If the centre of gravity lies on each of these lines, then there must be one point where all these lines meet. To find it, join each vertex to the midpoint of the opposite side, and the three lines *concur*, at the centre of gravity (Fig. 1.17). We have a bonus in this case, also. Because we are confident that the triangle, regarded as a plane lamina, has a centre of gravity, we can be confident that the three medians do indeed concur. We do not even need to draw a diagram to check this fact, whereas we only discovered that the altitudes concurred with the aid of a drawing.

Fig. 1.16

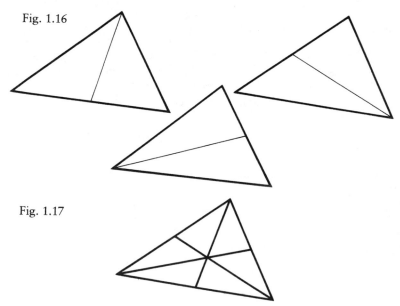

Fig. 1.17

Problem 1D   *Balancing a quadrilateral*

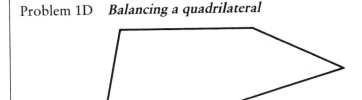

Fig. 1.18

Given that the centre of gravity of two objects taken together must lie on the line joining their individual centres of gravity, and given that the centre of gravity of any triangle can be found: where is the centre of the gravity of the quadrilateral shown in Fig. 1.18?

*

Already the triangle is looking less naked. We have two significant points, and two sets of three significant lines each. What other promising questions might we ask?

The very names, ortho*centre* and *centre* of gravity suggest a more general question: where is the centre of a triangle? The centre of an equilateral triangle, all of whose sides are equal, is an obvious point, because the triangle is so symmetrical, but what does it mean to talk about the centre of an unsymmetrical shape? The orthocentre (despite its name) can hardly qualify, because the orthocentre of a triangle can actually be outside the triangle, as in Fig. 1.19, which seems an

Fig. 1.19

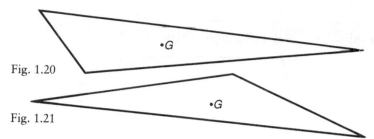

Fig. 1.20

Fig. 1.21

absurd place for a 'centre' to be. The centre of gravity might seem a better bet, but it is noticeable that in a long thin triangle, with two sides of roughly equal length, the centre of gravity is about twice as far from one vertex as from the other two (Fig. 1.20), while in a 'flat' triangle the centre of gravity is far closer to one vertex than either of the others (Fig. 1.21). These possibilities might be consistent with our idea of a 'centre', but on the other hand, they might not.

One answer for a typical triangle might be to draw a circle through its three vertices and call the centre of the circle, which will be equidistant from all three vertices, the centre of the triangle. Such a circle always exists and there is only one such circle, as we shall see. How can we draw this unique circle? By looking carefully, we could pick a point, round about X, which roughly fits this description, and by trial and error we could fiddle around until we had found the point more or less 'exactly' (Fig. 1.22). We can find it more quickly and accurately by considering that all the points that are the same distance from two given points lie on a straight line which bisects the line joining them at right angles. In the triangle, all the points which

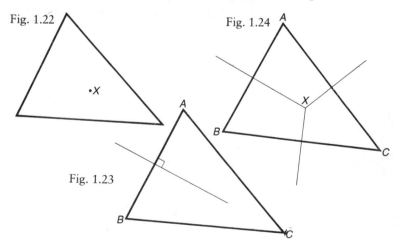

Fig. 1.22

Fig. 1.24

Fig. 1.23

are equal distances from A and B will lie on the thin line (Fig. 1.23), all the points equidistant from B and C on another, and the equidistant points from C and A on a third line. Since the circle we are seeking has only one centre, all three lines must pass through one point, the centre of the circle. In fact, if X is where the first two circles meet, then $XA = XB$ and also $XB = XC$, so $XA = XC$ and the third line also goes through X, which is therefore the centre of the unique circle through A, B and C (Fig. 1.24). For the third time we have a point of concurrency, the *circumcentre* of the triangle, and for the first time we have a totally convincing argument that the three lines really do concur.

Problem 1E *Interesting circles*

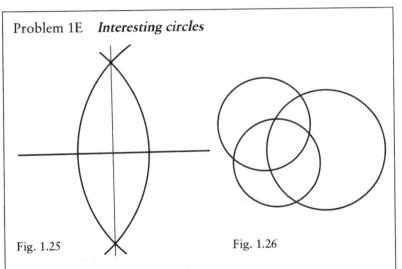

Fig. 1.25   Fig. 1.26

To find the perpendicular bisector of a straight line segment, it is sufficient to draw two circles (in practice, parts of circles will do) with their centres at the ends of the segment, and each with the same radius; see Fig. 1.25. The line joining their two intersections (provided that the circles are large enough to intersect) will be the perpendicular bisector of the line segment. Since the three vertices of a triangle can be chosen anywhere at all, this suggests that if we take any three circles, provided that they are the same size, then the three lines through their pairs of intersections will concur, at the point which is the circumcentre of the triangle formed by their centres.

What happens if, as in Fig. 1.26, the circles are *not* the same size?

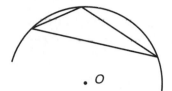

Fig. 1.27

Unfortunately for our plan, it turns out the circumcentre of an obtuse-angled triangle, like its orthocentre, lies outside the triangle! This is shown in Fig. 1.27.

Let us make one last attempt. Just as there is one circle through three points, there is one circle inside the triangle touching all three sides. The centre will be the same distance from each side. All the points equidistant from two given lines lie on the line which bisects them. So we have three lines, one for each vertex, and sure enough they meet at the centre of the *inscribed* circle (Fig. 1.28). Once again, if we are wide awake, there is a bonus waiting here. When two lines cross at a vertex, they form two angles, which have two angle bisectors, perpendicular to each other: one internal and one external.

The external angle bisectors concur in pairs with one of the internal bisectors at the centres of the three circles that touch two *produced* sides of the triangle, beyond the original triangle, and the third (original) side (Fig. 1.29).

Fig. 1.28

Fig. 1.29

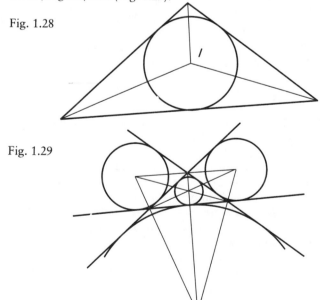

Problem 1F   *A familiar figure*

Fig. 1.30

We get an added bonus from Fig. 1.30, which shows the incentre and the three excentres, because the pairs of angle bisectors are perpendicular. What is familiar about this figure?

Unfortunately, the incentre is not a completely convincing candidate for *the* centre of a general irregular triangle, for familiar reasons. Is, perhaps, the very idea of a unique 'centre' of an irregular triangle a mirage? Indeed, it is. Different so-called 'centres' have different properties, and are useful for different purposes. This does not make the question 'What is the centre of a triangle?' useless – far from it: the very fact that the question was (frankly) vague has forced us to consider a variety of possibilities, each of which has its own points of interest. By exploring different possible answers, we have discovered a great deal about triangles in general.

The 'centres' we have discovered were all based on different ideas, but it does not follow that they are all unconnected to each other. If we take *any* mathematical object at all – a cube? a sphere? the number 144? a granny knot? – and make a list of its known features, there is a high chance that many of these features will be connected to each other.

Needless to say, the search for these connections is another way of finding out yet more about the object. The simplest way to look for connections between points and lines is to do a physical experiment, and draw them, as for example a typical irregular triangle drawn in Fig. 1.33, showing the orthocentre, the centre of gravity, the circum-

Problem 1G   *The shortest network*

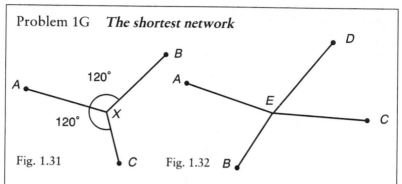

Fig. 1.31   Fig. 1.32

An engineer wishing to construct roads joining three towns, *A*, *B* and *C*, shown in Fig. 1.31, might naturally expect to minimize costs by making the total length of the roads a minimum. For reasons that will be revealed in Chapter 12, the ideal network (provided that the angles of the triangle are all less than 120° – an important condition) is joined at *X*, where the angles between the three new roads required are all 120°. With this information at hand, how can the total length of the network shown in Fig. 1.32, joining the *four* towns *A*, *B*, *C* and *D*, be reduced to a minimum in length? (You are not limited to using only four straight roads. *E* is merely where the roads cross; it has no significance otherwise.)

centre and incentre. A mere glance at these four points suggests that three of them lie in a straight line. Using a ruler, it appears also that $HG = 2 \times GO$. Repeated trials with other triangles will confirm this experimentally. The centre of gravity, the orthocentre and the circumcentre have been proved to lie on a line, called the Euler line, after the great eighteenth-century Swiss mathematician Leonhard Euler.

Fig. 1.33

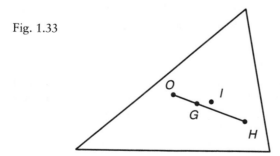

How does the mathematician *feel* on discovering that these already-special points are connected among themselves? The mathematician naturally accepts these new relationships as a sign of progress. Mathematicians are continually discovering odd and idiosyncratic facts about mathematical objects, but they are also continually finding connections between them, and they get a kick from the weird facts and the connections between them.

If all the facts of mathematics were isolated facts, they would have the attractions of novelty and surprise, but nothing more. If they were all connected together in obvious and trivial ways, they would excite no one. Because their connections are so often subtle and difficult to find, they are fascinating, and challenging. The history of mathematics is a long story of discovering novel objects and facts, and making sense of them by connecting them together into striking patterns.

Problem 1H   *The nine-point circle*

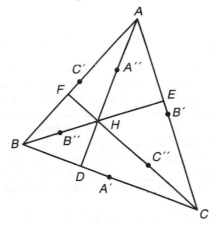

Fig. 1.34

On the triangle shown in Fig. 1.34 have been marked the altitudes, meeting at the orthocentre H; the midpoints of the sides, A', B' and C'; the midpoints A'', B'' and C'' of the segments AH, BH and CH; and the feet D, E and F of the altitudes. Why do the four points $B'C''B''C'$ form a rectangle? If a circle is drawn through the vertices of this rectangle, which other points in the figure will lie on it? (It will help to bear in mind that the line joining the midpoints of two sides of a triangle is parallel to the third side and one half of its length.)

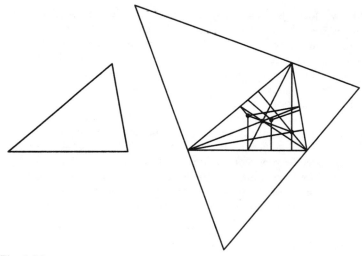

Fig. 1.35

Just as a sculptor taking a hammer and chisel to a piece of marble can see a vast number of potential statues within the marble, so the mathematician expects to find an endless variety of theorems in the simple shape of a triangle. In Fig. 1.35 is the original triangle side by side with another figure on which we have marked all the significant points and lines that we have discovered. How that empty space has filled up! So much so, that the result looks almost incomprehensible – we can only make sense of it a bit at a time. Yet we have looked at just a single triangle. What happens if we consider several triangles together? If we take several random triangles it is not impossible, but not so likely either, that there will be simple relationships between them.

However, if we take some triangles that already have a connection, then we can expect to find more connections – at least, that is how mathematicians have learnt to think from experience. Figure 1.36 shows four random straight lines, which form four triangles by taking the lines three at a time. These triangles are certainly connected because, if you draw any two of them, the other two are also determined. Each triangle has a unique circumcircle. What is the connection between these four circumcircles? We cannot help noticing that they all pass through a common point, O. If we look a little harder, we may also realize that O and the centres of the four circles all lie on a fifth circle, drawn thin in Fig. 1.36.

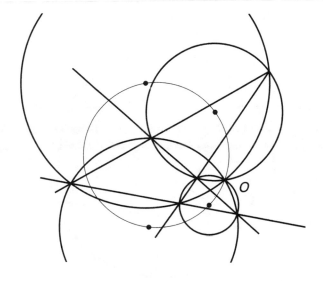

Fig. 1.36

What would happen if we had started with five lines? The five lines shown in Fig. 1.37 form five sets of four, by leaving out one line at a time. Each set of four has a special circle, as already described. What property is shared by all five sets of four lines?

Fig. 1.37

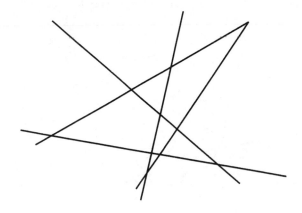

Problem 1I   *Four incentres*

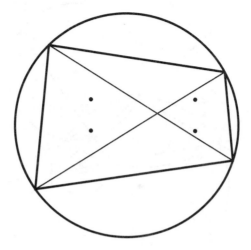

Fig. 1.38

Figure 1.38 shows four points on a circle joined to form four overlapping triangles. The four dots are the incentres on these triangles. By experiment, either judging by eye or by measuring, what properties does this figure have?

It is no great surprise to discover that the five circles, one for each set of four lines, all pass through a common point Q, and that the centres of the five circles all lie on another circle (Fig. 1.39).

Fig. 1.39

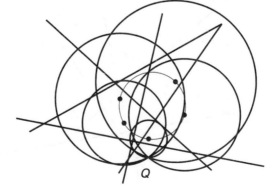

We might conjecture – correctly as it happens – that, if we had started with six lines, and taken them five at a time ... then we would have obtained six special circles ... and so on. This is just the start of an infinite chain of theorems.

Our next example is much simpler. These two triangles are similar in shape, and matching sides are parallel (Fig. 1.40). Just to glance at them, they almost look as if they were in perspective (partly because we placed the smaller triangle slightly higher and to one side). Draw the lines through matching vertices and these lines all meet in a point – three more concurring lines! The impression of perspective is now even stronger, and we can imagine the large triangle turning into the smaller triangle by moving towards it, getting smaller all the time. When the large triangle is half way towards the smaller, in the thin position (Fig. 1.41), its sides are exactly half way in length between

Fig. 1.40

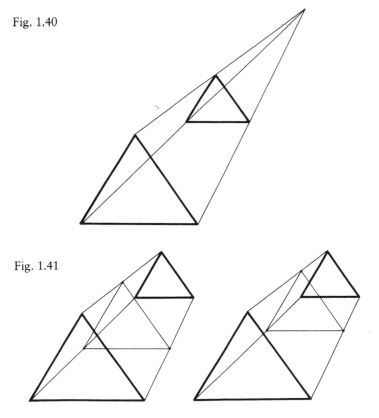

Fig. 1.41

Problem 1J *Ceva's theorem*

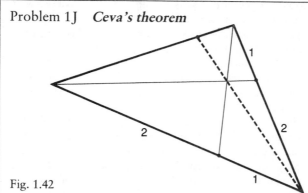

Fig. 1.42

To find the centre of gravity of a triangle, we joined the midpoints of the sides to the opposite vertices, and the resulting three lines concurred. In Fig. 1.42, two lines have been drawn from two vertices to points which divide the opposite sides in the ratio 2:1. If the third, dotted, line is added, in what ratio will it divide the third side? What is the general rule?

the original large triangle and the small triangle. Go two-thirds of the way from the larger to the smaller, and we get a new triangle of the same shape, whose sides in length will be two-thirds of the way from the sides of the larger triangle to the smaller.

This connection between triangles of the same shape, in the same orientation, seems so natural because it is so like perspective in the real world. It would seem just as natural that if the triangles were not in perspective, then all these features would be lost. Surprisingly, this is not so. In Fig. 1.43, the original two triangles are the same shape,

Fig. 1.43

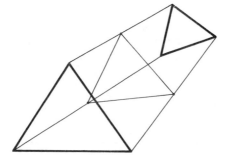

but in different orientations. We join corresponding vertices as before, and then mark the midpoints of these lines. What triangle do we get?

The perspective effect has indeed completely vanished, but the new triangle is still the same shape as the original triangles, although its sides are no longer the average of their sides. How does the mathematician feel now? Disappointed that the 'perspective' argument turns out to be misleading? No, not really, because the discovery that the 'wonky' triangles still produce another triangle of the same shape is itself so surprising that it more than makes up for any disappointment.

⁕

Problem 1K   *Putting the pebble in the box*

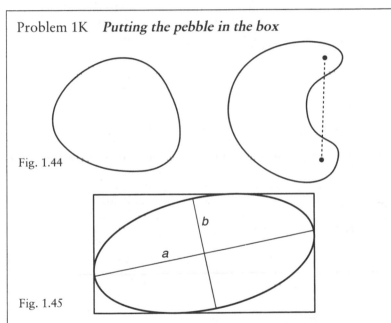

Fig. 1.44

Fig. 1.45

A convex region is represented on the left of Fig. 1.44. The region on the right is not convex, because some straight lines joining interior pairs of points go outside it, as the dotted line illustrates. There is a theorem that says that, if a region is convex, then it can be surrounded by a rectangle whose area is at most double the area of the region. For example, in Fig. 1.45, the ellipse of area πab has been surrounded by a rectangle whose area is approximately $4ab$, which is less than $2 \times \pi ab$. What is the connection between this theorem and triangles?

We started with a triangle in empty space with almost no visible features. We end up, after quite a short journey, with triangles, many special points, special lines, circles ... We have also, along the way, given names to many of these: 'altitudes', 'medians', 'circumcircles', 'incircles', 'orthocentre'. The very simple and even austere world in which we started is now populated with a large number of objects, with striking and surprising properties, and connections between them. It is, in fact, much more like the real world!

However, we have hardly started to explore the world of triangles. They will appear again in later chapters, when we discuss Pythagoras' theorem, Fermat points, triangles of forces, Morley's and Napoleon's theorems, triangular numbers, and barycentres. Meanwhile we conclude our brief foray with a problem (1K on page 21) which on the face of it has nothing to do with triangles; yet there is a connection.

Solutions to problems, Chapter 1

1A: *Across the river*

In Fig. 1.46, the man has adjusted the brim of his hat so that his eyes just see past the brim of the hat to the distant point X on the other bank of the river. With the ground, this forms a triangle whose shape is fixed by angle at which the brim of his hat is tipped. By turning his head, and looking past the brim of his hat to a point on the ground which is on his side of the river, he will be identifying a point he can reach on foot, which is the required distance away – and all he has to do is pace out the distance.

Fig. 1.46

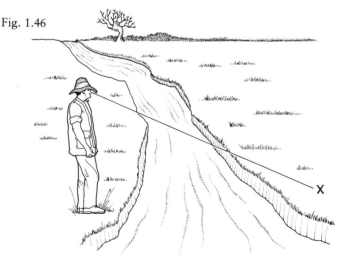

1B: *Bars and joints*

One joint is added with every pair of bars. So the number of bars is double the number of joints, with a small allowance for the initial situation in which one bar has a joint at either end. Therefore

$$\text{number of bars} = 2 \times \text{number of joints} - ?$$

By checking against any actual arrangement, the exact formula is established:

$$\text{number of bars} = 2 \times \text{number of joints} - 3.$$

1C: *Polygon angles*

If all the three quadrilaterals in Fig. 1.47 have the same angle sum, call it S, then the total angle sum of the two inner quadrilaterals is S more than the angle sum of the outer quadrilateral. But the difference between the total angle sum of the two inner and one outer quadrilaterals is just the starred angles, whose total sum is $2 \times 180° = 360°$. When the quadrilateral in Fig. 1.48 is cut by the line L, a triangle is removed and the remaining figure is a pentagon. The angle sum is decreased by the loss of angle A, but increased by the angles X and Y. Since $A + B + C = 180°$ and $B + X = 180°$ and $C + Y = 180°$, by adding the last two statements and subtracting the first, it follows that $X + Y - A = 180°$. Therefore, by adding one side, the angle sum has been increased by $180°$, and this will be true however many sides the original polygon had.

Fig. 1.47   Fig. 1.48

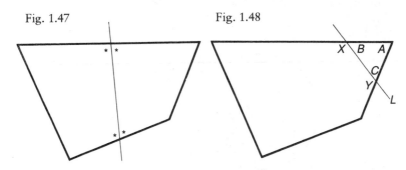

1D: *Balancing a quadrilateral*

Mark the centres of gravity of one pair of non-overlapping triangles, X and Y, and join them. The centre of gravity of the entire figure lies somewhere on the line shown in Fig. 1.49. Join the centres of gravity,

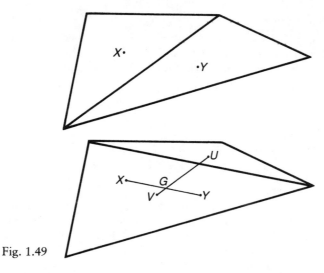

Fig. 1.49

U and V, of the other pair of triangles, and the same is true; so the centre of gravity, G, is where they intersect.

1E: *Interesting circles*

By experiment, if three circles of any size are chosen so that each circle intersects the other two, then the lines through the points of intersection concur. The point, O, where they meet, has a special property: $OA \times OA' = OB \times OB' = OC \times OC'$.

Fig. 1.50

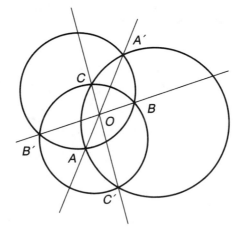

1F: *A familiar figure*

Since the pairs of angle bisectors are perpendicular to each other, the three internal and three external angle bisectors form a triangle with its altitudes. In other words, the incentre and three excentres form four orthocentric points.

1G: *The shortest network*

We can think of D, C and E as three vertices of a triangle (with angles all less than 120°) and DE and CE as the roads joining these three points, as in Fig. 1.51. But in that case we know that the shortest network joining them will be as in Fig. 1.52, where the roads all at angles of 120°. So we can reduce the length of the network by replacing DE and CE by the dotted lines as shown. However, we can now apply much the same argument to the lines AE, BE and the dotted line to E. The network can be reduced in length still further if we replace them with three straight lines meeting at 120°. This is a marked improvement on the original four lines meeting at E. It is not yet, however, the best possible solution, because by moving E to the left, we have slightly changed the angles at which the dotted roads meet – they are no longer all 120°! So we have to make some more slight adjustments . . .

Fig. 1.51

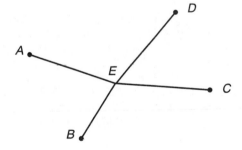

Fig. 1.52

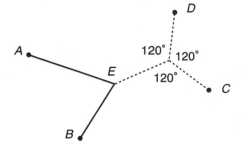

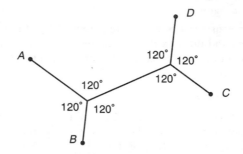

Fig. 1.53

The final result, when no more improving adjustments can be made, is the network shown in Fig. 1.53. One final question remains: what would have happened if we had started by considering, for example, triangle *AED* in the original figure, instead of triangle *DEC*? Answer – nothing, because the angles *AED* and *BEC* are both greater than 120°, and so the lines *AE* and *ED* already make the shortest road network joining *A*, *E* and *D*, when *B* and *C* are ignored. To improve the overall network, we must start with one of the triangles *DEC* or *AEB*.

1H: *The nine-point circle*

In Fig. 1.34, *B′C′* is ½*BC* (from the triangle *ABC*) and also *B″C″* = ½*BC* (from the triangle *HBC*). Also, *B′B″* and *C′C″* are each one half of *AH*, and parallel to *AH*, from the triangles *AHB* and *AHC*. Since *AH* is perpendicular to *BC*, *B′C″B″C′* is a rectangle, and so also is *B′A″B″A′*.

If a circle is drawn through *B′C″B″C′*, then *B′B″* will be a diameter of this circle. However, *B′B″* is also a diameter of the circle through the corners of the rectangle *B′A″B″A′*. Therefore the same circle passes through the vertices of both those rectangles (and the rectangle *A′C′A″C″*), and so all six points *A′*, *B′*, *C′*, *A″*, *B″*, *C″* lie on the same circle.

These six points can also be described as the midpoints of the six line segments joining the four orthocentric points. If this circle is drawn exactly, then the figure will show that the feet of the three altitudes also lie on this circle, so it is actually a nine-point circle. This follows because *A′DA″*, *B′EB″* and *C′FC″* are all right angles.

1I: *Four incentres*

By measurement, Fig. 1.54 has all the following properties. The four incentres, A, B, C and D, are the vertices of a rectangle. The lines, PQ and RS, which bisect the rectangle through the midpoints of opposite pairs of sides, meet the circle at P, Q, R and S which are the midpoints of the arcs XY, ZU, YZ and UX. In addition, these midpoints are in line with the incentres and the vertices of the original quadrilateral. For example, XBR is a straight line.

Fig. 1.54

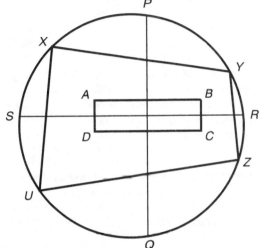

1J: *Ceva's theorem*

In Fig. 1.55, BP/PC = 2/1 and CQ/QA = 2/1. By measurement, AR/RB = 1/4. This is a case of Ceva's theorem, named after Giovanni

Fig. 1.55

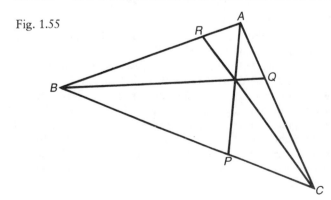

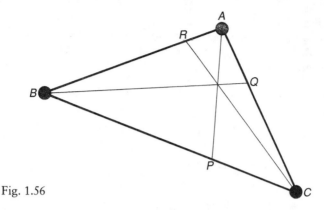

Fig. 1.56

Ceva (c. 1647–1736), which says that if three lines through the vertices of a triangle concur, then $AR/RB \times BP/PC \times CQ/QA = 1$. This can be 'proved' by analogy, by thinking of three *different* weights w_A, w_B, w_C, one at each of the vertices of the triangle (see Fig. 1.56). We can choose the three weights so that $w_B/w_C = CP/PB$ which means that P will be the centre of gravity of the two weights at B and C, and the centre of gravity of all three weights will then lie on the line PA. In Fig. 1.56, we have placed 1 unit at B and 2 units at C and 4 units at A. This makes P the centre of gravity of the weights at B and C, and it also makes Q the centre of gravity of the weights at C and A, so that the centre of gravity of all three weights lies on the line QB.

The centre of gravity of all three weights is therefore where AP and QB meet, and the centre of gravity of A and B must lie on the line CR; in fact it must be at R, which therefore divides AB in the ratio $1:4$.

1K: *Putting the pebble in the box*

If the curve bounding the convex region is anything like a circle or an ellipse, then the area of the rectangle will be much less than double the area of the region. To make the ratio anywhere near double, we need a curve with three 'vertices' and rather 'flat' curved sides joining them. However, we can go further. The definition of a convex region does not say that its bounding curve must in fact be curved! It only says that the region will not be convex if you can join two interior points by a straight line which goes outside the region. But for an ordinary triangle this cannot be done. Therefore a triangle counts, according to the definition, as a convex region, and we can take the triangle in the figure as an extreme case of the previous 'curved' triangle. In this extreme case, the rectangle is exactly double the area of the triangle.

Fig. 1.57

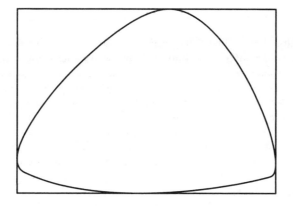

Fig. 1.58

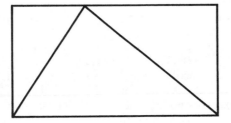

Chapter 2
Numbers and patterns

A headteacher, asked what pupils arriving for their first day in secondary school should know about arithmetic, replied that they should be on friendly terms with the numbers from 1 to 100. How can you be on 'friendly terms' with numbers? At the very least it suggests that you do not panic and start sweating at the sight of a sum, but more than that it implies an easy familiarity with them. You have met them before, you know some of their properties, you know at least some of the many relationships between them. You know quite a few facts about them, and maybe you have discovered some of these facts yourself.

Many people keep a small store of number facts in their heads. A gross, 144, is a dozen dozen. The numbers 169 and 196 are both squares, of 13 and 14 respectively. You might remember that 16, 32, 48 and 64 used to be the numbers of ounces in 1 to 4 pounds weight. Now that we are decimalized, 32, 64, 128, 256 and 512 are better known as powers of 2; computer buffs recognize these figures because they turn up in the descriptions of computer memories and in the labels of computers.

The Indian genius Ramanujan was familiar with many more numbers, as G. H. Hardy discovered when he visited him in hospital. Hardy remarked that he had travelled in Taxi No. 1729, adding that the number seemed to him a rather dull one, and he hoped that this was not an unfavourable omen. Ramanujan replied at once, 'No, it is a very interesting number, it is the smallest number expressible as the sum of two cubes in two different ways.' In fact, $1729 = 12^3 + 1^3 = 10^3 + 9^3$.

Anyone who works with numbers naturally acquires a stock of useful facts, as well as some apparently useless ones. Anyone might notice that the last single-digit number is 9, which is a perfect square, 3×3. Is this deeply significant? No, it is more of an accident. However, you might also notice that numbers which are one less than perfect squares are the products of two whole numbers which differ by 2. For examle, $16 - 1 = 15$, and 15 is 3×5. Similarly, $9 - 1 = 8$ and $8 = 2 \times 4$, and also $36 - 1 = 35$ and $35 = 5 \times 7$.

These facts are more naturally appealing, because they seem to reveal an underlying *pattern*. It is just so tempting to suppose that this property applies to any square at all, and the moment we start to wonder whether it does, then we are thinking mathematically. In fact, there are any number of patterns in the whole numbers, from simple patterns like this one to patterns so complex and subtle that the greatest mathematicians have never been able to prove that the pattern does in fact go on for ever, or even that it exists at all.

The sequence of integers starting from 1 already has one strong pattern – the unit digits (at the right-hand end) repeat in the sequence 1 2 3 4 5 6 7 8 9 0 1 2 3 4 5 . . . and the 'tens' digits repeat in the pattern (0) 1 2 3 4 5 6 7 8 9 0 . . . and so on. Higher-place values repeat in the same pattern, which is why even children can do today what the Greeks and Romans found difficult: count to very high numbers by following a simple pattern.

However, supposing only that we can add, subtract, multiply and divide, we can find many more patterns than this. Following the example of Chapter 1, we can hope to find special numbers, and special sets of numbers, which are outstanding for their remarkable and curious properties.

The perfect squares are among the most familiar numbers:

$$1 \quad 4 \quad 9 \quad 16 \quad 25 \quad 36 \quad 49 \quad 64 \quad 81 \quad 100 \quad \cdots \quad .$$

We can hardly avoid spotting that the differences between them get steadily larger:

$$1 \quad 4 \quad 9 \quad 16 \quad 25 \quad 36 \quad 49 \quad 64 \quad 81 \quad 100 \quad \cdots$$

$$3 \quad 5 \quad 7 \quad 9 \quad 11 \quad 13 \quad 15 \quad 17 \quad 19 \quad \cdots \quad .$$

Moreover, the sequence of differences is very simple – it is just the sequence of odd numbers. This suggests the question: 'Is the sequence of differences always simpler than the original sequence?'

We can test this idea against the powers of 2 already mentioned:

$$2 \quad 4 \quad 8 \quad 16 \quad 32 \quad 64 \quad 128 \quad 256 \quad \cdots \quad .$$

Each power is double the power to its left. Alternatively, working backwards, each number is one half of the number to its right, which suggests that perhaps we should halve the initial 2 and put a 1 at the start of the sequence.

$$1 \quad 2 \quad 4 \quad 8 \quad 16 \quad 32 \quad 64 \quad 128 \quad 256 \quad \cdots \quad .$$

When we write down the differences,

$$1 \quad 2 \quad 4 \quad 8 \quad 16 \quad 32 \quad 64 \quad 128 \quad 256 \quad \cdots$$
$$1 \quad 2 \quad 4 \quad 8 \quad 16 \quad 32 \quad 64 \quad 128 \qquad \cdots \quad ,$$

we notice that the same sequence appears repeated, exactly. So it seems that this is quite a different *type* of sequence from the sequence of squares, and also that the answer to our question is 'no'.

The number 64 is also a cube: $64 = 4 \times 4 \times 4$. The sequence of cubes starts as

$$1 \quad 8 \quad 27 \quad 64 \quad 125 \quad 216 \quad 343 \quad \cdots$$

and increases more rapidly than the squares. We can check just how fast, as before, by writing down the differences between one cube and the next:

$$1 \quad 8 \quad 27 \quad 64 \quad 125 \quad 216 \quad 343 \quad \cdots$$

$$7 \quad 19 \quad 37 \quad 61 \quad 91 \quad 127 \quad \cdots$$

$$12 \quad 18 \quad 24 \quad 30 \quad 36 \quad \cdots \quad .$$

The last line is the differences between the differences. It is also rising steadily, but not so quickly, just 6 at a time.

With the number 6 temporarily on our minds, we are more likely to spot that the first line of differences can be written,

$$1 \qquad 8 \qquad 27 \qquad 64 \qquad 125 \qquad \cdots$$
$$1 \times 6 + 1 \quad 3 \times 6 + 1 \quad 6 \times 6 + 1 \quad 10 \times 6 + 1 \quad \cdots$$

The mathematician will naturally suspect that there is something special about the sequence of multipliers of 6:

$$1 \quad 3 \quad 6 \quad 10 \quad 15 \quad 21 \quad 28 \quad 36 \quad 45 \quad \cdots$$

(we have added a few more for good measure). Is there? Mathematicians know from experience that a pattern which turns up in the *middle* of a problem is especially likely to be significant. There is something about problems – one problem leads to another, and one good idea leads to several other good ideas.

What patterns are there here? The pattern of *differences* is very simple:

$$1 \quad 3 \quad 6 \quad 10 \quad 15 \quad 21 \quad 28 \quad 36 \quad 45 \quad \cdots$$

$$2 \quad 3 \quad 4 \quad 5 \quad 6 \quad 7 \quad 8 \quad 9 \quad \cdots$$

Problem 2A *Sums of powers of 2*

Although we could say that 'there are not many powers of 2' and certainly almost all numbers are *not* powers of 2, there are just enough powers of 2 at the beginning of the sequence to make the small odd numbers by addition.

Thus: $3 = 1 + 2$ $5 = 4 + 1$ $7 = 4 + 2 + 1$

We can also make the missing even mumbers: $6 = 4 + 2$ and $10 = 8 + 2$.

How many integers can be made by adding up powers of 2 without using any power twice? In how many ways can an integer, such as 85, be made by adding up different powers of 2?

The second differences are just the sequence of integers, but with 1 missing. This suggests to the mathematician that the sequence of differences should start with a 1, which it would do if the sequence of cubes had started with $0^3 = 0$. So let's add that:

0	1	8	27	64	125 ...

$0 \times 6 + 1$ $1 \times 6 + 1$ $3 \times 6 + 1$ $6 \times 6 + 1$ $10 \times 6 + 1$ ...

So our sequence becomes

$$0 \quad 1 \quad 3 \quad 6 \quad 10 \quad 15 \quad 21 \quad \cdots$$
$$1 \quad 2 \quad 3 \quad 4 \quad 5 \quad 6 \quad \cdots$$

Now everything looks neat and complete and the mathematician is satisfied, and can look for more patterns in the sequence.

As it happens, $1 + 3 = 4$ which is 2^2, and $3 + 6 = 9$ which is 3^2. This is no coincidence: the sum of any adjacent pair is a perfect square – which is rather curious because this sequence turned up 'out of the blue' while we were looking at the sequence of *cubes*, not squares.

You might also spot that the numbers in the sequence all have nice easy factors: we can write the sequence as:

1	3	6	10	15	21	28	36	45
1×1	1×3	2×3	2×5	3×5	3×7	4×7	4×9	5×9

The pattern is very strong. The even numbers and the odd numbers occur in overlapping pairs, as it were. This pattern can be made even stronger, and smoother, by writing it like this:

$$\tfrac{1}{2} \times 1 \times 2 \quad \tfrac{1}{2} \times 2 \times 3 \quad \tfrac{1}{2} \times 3 \times 4 \quad \tfrac{1}{2} \times 4 \times 5$$
$$\tfrac{1}{2} \times 5 \times 6 \quad \tfrac{1}{2} \times 6 \times 7 \quad \tfrac{1}{2} \times 7 \times 8 \quad \tfrac{1}{2} \times 8 \times 9$$

This pattern in turn might suggest a geometrical picture of these numbers, by drawing rectangles with sides differing by one unit, and dividing them into two (see Fig. 2.1). Each number on the sequence is one half of a rectangle, so the nth triangular number is $\tfrac{1}{2}n(n + 1)$. As it happens – but this is hardly a coincidence! – Fig. 2.2 explains why pairs of adjacent triangular numbers make the squares.

If we turn the triangles round (Fig. 2.3), it also explains why the differences between them increase one at a time. In fact, these numbers were called the triangular numbers by the ancient Greeks,

Fig. 2.1

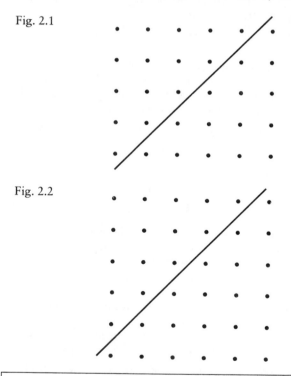

Fig. 2.2

Problem 2B *Triangular numbers and large squares*

The same kind of dot diagram can be used to show that, if T is a triangular number, then $8T + 1$ must be a perfect square. How?

Fig. 2.3

just because they were formed by making triangles of dots, or pebbles. We did not discover them like that, of course, but by seeing how quickly the sequence of cubes increased. This is quite typical of the way mathematics works. One thing leads to another, and everything seems to be connected to everything else.

The triangular numbers have many curious properties. Here are two of them. (For convenience, we will denote the sequence of nonzero triangular numbers by T_1, T_2, T_3 and so on, so $T_n = \tfrac{1}{2}n(n + 1)$.) The sums of powers of 9, starting with 1, are all triangular numbers:

$$
\begin{aligned}
1 &= 1 &&= T_1 \\
1 + 9 &= 10 &&= T_4 \\
1 + 9 + 9^2 &= 91 &&= T_{13} \\
1 + 9 + 9^2 + 9^3 &= 820 &&= T_{40}
\end{aligned}
$$

$\vdots$

Problem 2C *Triangular numbers and cubes*

There is another connection between the triangular numbers and the cubes, which is to do with the squares of the triangular numbers:

triangulars	1	3	6	10	15	21	$\cdots$
their squares	1	9	36	100	225	441	$\cdots$

What is it?

Problem 2D *Triangular numbers and squares*

There is also a strong connection between the square of a triangular number and the triangular numbers on either side of it:

triangulars   1   3   6   10   15   21 $\cdots$

36

What is it?

Problem 2E *Sums of triangular sequences*

You might notice that the sum of the first three triangular numbers equals the fourth:

$$1 + 3 + 6 = 10$$

What other sums of consecutive triangular numbers are equal?

A rather different property: every perfect fourth power is the sum of two triangular numbers, according to a simple pattern:

$$2^4 = \quad 16 = \quad 1 + \quad 15 = T_1 + T_5$$
$$3^4 = \quad 81 = \quad 15 + \quad 66 = T_5 + T_{11}$$
$$4^4 = \quad 256 = \quad 66 + \quad 190 = T_{11} + T_{19}$$
$$5^4 = \quad 625 = \quad 190 + \quad 435 = T_{19} + T_{29}$$
$$6^4 = 1296 = \quad 435 + \quad 861 = T_{29} + T_{41}$$
$$7^4 = 2401 = \quad 861 + 1540 = T_{41} + T_{55}$$
$$\vdots$$

The mere fact that all fourth powers are the sum of only two triangular numbers is not quite as surprising as it might seem. Fermat claimed that every integer is the sum of at most three triangular numbers. In fact he went further. The Greeks had defined not just triangular numbers but also square, pentagonal, hexagonal . . . numbers by patterns of pebbles. We know the triangular and square numbers. Figure 2.4 shows the beginnings of the sequences of penta-

Fig. 2.4

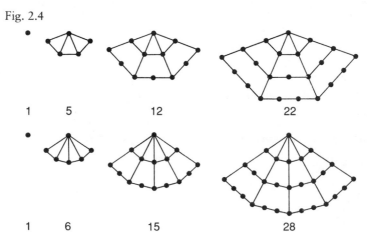

1　　5　　　　12　　　　　　22

1　　6　　　　15　　　　　　28

gonal and hexagonal numbers. Pierre de Fermat (1601–65) made the remarkable but true claim that each integer is the sum of at most three triangular numbers, four square numbers, five pentagonals, six hexagonals, and so on.

Fermat was a lawyer, and parliamentarian, and one of the greatest amateur mathematicians of all time. Among his many achievements, he invented coordinate geometry, at the same time as Descartes. He explained how light is refracted when it passes from air to water, by assuming that 'nature operates by the simplest and most expeditious ways and means', meaning that a ray of light always took the shortest path (in terms of time) between two points. (We shall see two everyday examples of the same idea in Chapter 7, p. 207.)

Fermat also investigated numbers that are the sum of two squares. An example of such a number is 25 ($= 3^2 + 4^2$), which happens itself to be a square ($25 = 5^2$). The relation $3^2 + 4^2 = 5^2$ is familiar from the simplest illustration of Pythagoras' theorem (Fig. 2.5), and (3, 4, 5) is the simplest example of a *Pythagorean triple*.

Looking again at the sequence of squares, we might ask: which other squares are the sums of two squares?

1   4   9   16   25   36   49   64   81   100   121   144   169 $\cdots$.

There are not so many: the next smallest solution is

$$5^2 + 12^2 = 13^2.$$

After this comes $6^2 + 8^2 = 10^2$ which, however, is only the 3–4–5 solution with each side of the triangle doubled. Next comes $7^2 + 24^2 = 25^2$. The calculation is now becoming tedious (though a

Fig. 2.5

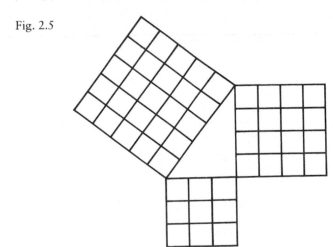

computer would help), but there is a pattern to aid us: in the 3–4–5 triangle, $4 + 5 = 9 = 3^2$, and in the 5–12–13 triangle $12 + 13 = 25 = 5^2$. This also fits the 7–24–25 triangle, because $24 + 25 = 49 = 7^2$ and we can now, with a high degree of confidence *predict* that

$$9^2 + 40^2 = 41^2, \quad 11^2 + 60^2 = 61^2 \quad \text{and so on.}$$

Calculation confirms that this is correct, and we now have a method of writing down as many Pythagorean triples as we wish.

Going back to the trial-and-error testing that we might have used to discover these Pythagorean triples, there is another pattern that we might have spotted. When we add up pairs of squares in the hope of getting another square, we are usually disappointed. However, we do get a suspiciously large number of prime numbers – a *prime number* being a number having no factors apart from itself and 1, with the exception of the number 1 itself, which is classed as a nonprime.

This appearance of primes is surprising, because there are not that many of them, and they occur in a random sort of way, unlike the squares. The sequence of primes starts

$$\begin{array}{ccccccccccccc} 2 & 3 & 5 & 7 & 11 & 13 & 17 & 19 & 23 & 29 & 31 & 37 & 41 \\ 43 & 47 & 53 & 59 & 61 & 67 & 71 & 73 & 79 & 83 & 89 & 97 & \ldots \end{array}$$

Four of the first ten integers are prime, and one quarter of the first one hundred, and 168 of the first 1000 integers, or roughly one sixth. Thereafter the proportion of primes drops steadily.

Table 2.1 gives the results of adding pairs of different squares.

Problem 2F   *An impossible sum of squares*

Fermat claimed that every integer is the sum of four squares, but many numbers are the sum of fewer than four. For example, $6 = 4 + 1 + 1$, $10 = 9 + 1$ and $14 = 9 + 4 + 1$. However, 7 is certainly not the sum of only three squares.

Here is a statement about perfect squares: 'Every perfect square, when divided by 8, leaves a remainder of either 0, 1 or 4.' Why is this statement true, and how does it help to prove that none of the numbers

$$7 \quad 15 \quad 23 \quad 31 \quad 39 \quad 47 \quad \cdots \quad,$$

(extended by repeatedly adding 8) is the sum of only three squares?

Table 2.1

	1	4	9	16	25	36	49	64	81	100
1		**5**	10	**17**	26	**37**	50	65	82	**101**
4			**13**	20	**29**	40	**53**	68	85	104
9				25	34	45	58	**73**	90	**109**
16					**41**	52	65	80	**97**	116
25						**61**	74	**89**	106	125
36							85	100	117	136
49								**113**	130	**149**
64									145	164
81										**181**

Bearing in mind that alternate entries in the table must be even because they are sum of two even or two odd squares, we notice that the number of primes (printed in bold) is surprisingly large. Here they are in sequence. They include nearly half of all the primes up to 100, eleven out of twenty-five:

$$5 \quad 13 \quad 17 \quad 29 \quad 37 \quad 41 \quad 53 \quad 61 \quad 73 \quad 89 \quad 97$$

This might seem a slight cause for surprise, but there is a bigger surprise if we look at the factors of the numbers which are *not* prime. Table 2.2 is based on Table 2.1, but shows only the factors of the nonprime entries.

Amazingly, almost all the factors are either 2 or primes which occur themselves in the original table (Table 2.1). The only exceptions are the factors 3 which occur, naturally, when 3^2, 6^2 and 9^2 are added together.

Table 2.2

	1	4	9	16	25	36	49	64	81	100
1			2×5		2×13		$2^2\times5^2$	5×13	2×41	
4				$2^2\times5$		$2^3\times5$		$2^2\times17$	5×17	$2^3\times13$
9				5^2	2×17	$3^2\times5$	2×29		$2\times3^2\times5$	
16						$2^2\times13$	5×13	$2^4\times5$		$2^2\times29$
25							2×37		2×53	5^3
36							5×17	$2^2\times5^2$	$3^2\times13$	8×17
49									$2\times5\times13$	$2^2\times7^2$
64									5×29	$2^2\times41$

Problem 2G   *Sums and differences of squares*

It is a basic theorem of algebra that the difference of two squares is the product of their sum and their difference. In other words,

$$p^2 - q^2 = (p + q)(p - q).$$

For example, $7^2 - 5^2 = 24 = (7 + 5)(7 - 5)$. How does this fact explain the 5–12–13 and 7–24–25 pattern described in the text?

What is happening here? What is so special about these prime numbers? These two questions require two answers. The second is the simpler. The primes are the only even prime, 2, together with the primes which are 1 more than a multiple of 4. Thus $13 = 3 \times 4 + 1$. On the other hand, the prime $11 = 3 \times 4 - 1$ and is not included in the list.

As stated by Fermat, every prime that is 1 more than a multiple of 4 is the sum of two squares, in one way only, and *no* prime that is 1 less than a multiple of 4 is the sum of two squares.

Much earlier, Leonardo of Pisa (*c.* 1175–1220) knew that the product of two numbers, each of which is the sum of two squares, is also the sum of two squares. In fact, it is usually so in two ways; for example,

$$(2^2 + 5^2)(1^2 + 3^2) = 290 = 17^2 + 1^2 = 13^2 + 11^2.$$

The pattern which explains this looks like this:

$$(2^2 + 5^2)(1^2 + 3^2) = (1 \times 2 + 5 \times 3)^2 + (2 \times 3 - 5 \times 1)^2$$
$$= (1 \times 2 - 5 \times 3)^2 + (2 \times 3 + 5 \times 1)^2.$$

We could also put this in symbols, thus:

$$(a^2 + b^2)(c^2 + d^2) = (ac + bd)^2 + (ad - bc)^2$$
$$= (ac - bd)^2 + (ad + bc)^2.$$

We now have a much more powerful method of creating Pythagorean triples. First we take any number which is the sum of two squares, and square it. The result is a square which will itself be the sum of two squares. For example: $1^2 + 4^2 = 17$,

$$17^2 = (1^2 + 4^2)(1^2 + 4^2) = (1 \times 1 + 4 \times 4)^2 + (1 \times 4 - 4 \times 1)^2$$
$$= (1 \times 1 - 4 \times 4)^2 + (1 \times 4 + 4 \times 1)^2.$$

$$(*)$$

The first solution is not helpful, because the second square is zero, so we only discover what we already knew, that $17^2 = 17^2$. The second solution, however, tells us that $17^2 = 15^2 + 8^2$.

We can turn this pattern into algebra. Suppose that $z = m^2 + n^2$. Then, following the pattern of (∗) above, we have

$$z^2 = (m^2 + n^2)(m^2 + n^2) = (m^2 - n^2)^2 + (mn + mn)^2.$$

So

$$(m^2 + n^2)^2 = (m^2 - n^2)^2 + (2mn)^2.$$

The Babylonians about 2000 BC were familiar with 'Pythagorean triples', more than fifteen hundred years before Pythagoras. A famous cuneiform tablet, named Plimpton 322, lists fifteen right-angled triangles with integral sides, and it seems likely that the Babylonians knew that the three expressions $m^2 + n^2$, $m^2 - n^2$ and $2mn$ give the sides of a right-angled triangle.

Problem 2H   *A formula for Pythagorean triples*

Take any pair of consecutive odd numbers such as 7 and 9. (A pair of consecutive even numbers will do as well.) Add their reciprocals:

$$1/7 + 1/9 = 16/63.$$

Then 16 and 63 are two sides of a right-angled triangle. In fact,

$$16^2 + 63^2 = 65^2$$

Why does this operation work?

Problem 2I   *Sums of sequences of squares*

This pattern somewhat resembles the pattern of Problem 2E.

$$3^2 + 4^2 = 5^2$$

$$10^2 + 11^2 + 12^2 = 13^2 + 14^2$$

$$21^2 + 22^2 + 23^2 + 24^2 = 25^2 + 26^2 + 27^2$$

$$\vdots$$

What are the patterns in the sequence 3, 10, 21, . . . which starts these equations?

How can this pattern be explained by using 'differences of two squares'?

The last two problems might suggest that there is an endless supply of number patterns involving the squares, the triangular numbers, the cubes, . . . , and indeed this is the case. At first they seem very surprising, but after a while the simpler patterns become familiar, and easy to construct.

However, as usually happens in mathematics, once we get used to the simpler patterns, we find beyond them more complicated and more *powerful* patterns, based on more subtle ideas. Here is an example:

$$1 \quad 16 \quad 81 \quad 256 \quad 625 \quad 1296 \quad 2401 \quad 4096 \quad \ldots \quad .$$

This is the sequence of fourth powers, which we have already met. Looking at the last digits, a fact might strike us: with the exception of $5^4 = 625$, every other fourth power seems to be one more than a multiple of 5.

Is this a coincidence? No, it is not. Further checking with a calculator will confirm that all fourth powers, except the fourth powers of 5, 10, 15, and so on, leave remainder 1 when divided by 5.

Moreover, if we search for similar patterns with other powers, we find them for 6th powers, which leave remainder 1 when divided by 7 (with the exceptions we would expect) and 10th powers, which leave remainder 1 when divided by 11 (with the same exceptions).

These are all examples of Fermat's Little Theorem which says that, if p is a prime, then the remainder when n^{p-1} is divided by p is 1, provided that n is not a multiple of p. So, for example, 13^{7-1} leaves the remainder 1 when divided by 7.

Fermat's Little Theorem cannot be proved by the kind of arguments that prove the results about triangular numbers which we have seen earlier. It needs new ideas. We will sketch a proof with the aid of an example. You are warned that, compared to almost all the other arguments in this book, this argument is difficult!

Consider the sequence

$$3 \quad 3^2 \quad 3^3 \quad 3^4 \quad 3^5 \quad 3^6 \quad 3^7 \quad 3^8 \quad 3^9 \quad 3^{10}$$

or

$$3 \quad 9 \quad 27 \quad 81 \quad 243 \quad 729 \quad 2187 \quad 6561 \quad 19683 \quad 59049.$$

This is the sequence of powers of 3, up to 3^{10}. We are interested in the remainders when these are divided by the prime 11. We shall not keep on referring to the number 11, so please remember that all references to 'remainders' actually mean 'remainders after division by 11'. This is essential.

We start by thinking about the *smallest* power of 3 that leaves a remainder 1, assuming (as can be proved) that such a power exists. Call it 3^a. Next, consider the possibility that two powers of 3, say 3^x the larger and 3^y the smaller, *both less than* 3^a leave the same remainder. Then their difference will be a multiple of 11; so

11 will divide $3^x - 3^y$.

But in this case we can factorize 3^x and 3^y and draw the conclusion that

11 will divide $3^y(3^{x-y} - 1)$.

But this means, since 11 cannot divide the power of 3, that

11 divides $3^{x-y} - 1$

or

3^{x-y} leaves remainder 1 on division by 11.

But this is not possible, because we have already assumed that 3^a is the smallest power of 3 to leave remainder 1. It follows that all the powers of 3 up to 3^a leave different remainders. (The *existence* of a power of 3 leaving remainder 1 can be shown by reasoning that, since the eleven numbers $3^1, \ldots, 3^{11}$ can leave only the remainders 1 to 10, two of them, say 3^r and 3^s with $r < s$, leave the same remainder, and hence 3^{s-r} leaves remainder 1, by the factorization method above.)

The argument so far has been an example of Fermat's 'method of descent'. It was his favourite method which he used at every opportunity. The point of the method is to assume that something can be done, and that there is a *least number* for which it is possible. You then demonstrate that, if it is indeed possible, then there is a number *less than* that least number for which it is also possible. This is a contradiction which shows that your original assumption was false. This method can be extremely powerful, but it isn't easy.

That was the first part of the proof, and we can illustrate it by calculating the remainders in the last table, until we reach a remainder 1. It turns out that, in this case, the smallest power of 3 with remainder 1 is 3^5:

3	3^2	3^3	3^4	3^5	3^6	3^7	3^8	3^9	3^{10}
3	9	27	81	243	729	2187	6561	19683	59049
3	9	5	4	1.					

Our next step is to pick the smallest remainder that is not included in the sequence 3–9–5–4–1. It is 2. Now multiply 2 by 3, again and again, always writing down the remainder on division by 11, when the answer is over 11:

2	2×3	2×3^2	2×3^3	2×3^4	
2	6	18	54	162	
2	6	7	10	8	[2 6 7 10 8 2 6 7 ...]

The terms in brackets show how the sequence of remainders would repeat if we continued. We notice that the sequence 2–6–7–10–8 contains the same number of terms as the sequence 3–9–5–4–1. Why should this be so? Well, in the first place, since $3^5 = 243$ leaves remainder 1, we can be certain that 2×3^5 will leave remainder 2. In other words, after at most five steps the sequence will return to 2 and repeat.

Why could it not return to 2 in less than five steps? Because in that case, by much the same argument as before, we would be able to find a power of 3 less than 3^5 with remainder 1, and we have already determined that this is not possible.

So we see why the sequence starting with 2 is just as long as the original sequence. There is now one final step to take: we check to see if there are any other remainders which have not appeared in these two sequences. If there were, then we would take the smallest and do with it what we did with 2, to get another sequence of the same length. But, no, there are not. Every remainder is included in these two sequences. Further, the two sequences have no member in common (which we could prove by reasoning as we did to show that the first sequence had no repeated member).

With a pause to let the significance of these facts sink in, we now know what we wanted. The sequences 3–9–5–4–1 and 2–6–7–10–8 are the same length and they total 10. Therefore each of them is *one half* of 10 in length. That is what we wanted to know – that the length of 3–9–5–4–1 was a neat fraction of 10. It means that if we go back to our original table and continue to write down the remainders,

3	3^2	3^3	3^4	3^5	3^6	3^7	3^8	3^9	3^{10}
3	9	27	81	243	729	2187	6561	19683	59049
3	9	5	4	1	3	9	5	4	1,

the sequence 3–9–5–4–1 will fit in exactly, and the remainder 1 will appear under 3^{10}. In other words, 3^{10} leaves remainder 1 when divided by 11.

Solutions to problems, Chapter 2

2A: *Sums of powers of 2*

Take any integer and subtract from it the greatest power of 2 less than it. The remainder will be *less than* that power of 2 (because otherwise you could have subtracted the next greater power of 2). Now, in turn, subtract from the remainder the greatest power of 2 less than it, and repeat until the remainder is zero.

The result of this process is a *unique* way of representing the original integer as the sum of powers of 2. It works for every integer. Therefore the answers are 'all integers' and 'in exactly one way'.

Another way to deduce the same result is to take the original number and divide by 2 repeatedly, noting the sequence of remainders. An example will illustrate the process. For convenience, we start with the original number on the right and work backwards, noting the remainders underneath:

$$0 \quad 1 \quad 2 \quad 5 \quad 11 \quad 23 \quad 47 \quad 94 \quad 189 \quad 379 \quad 758$$
$$1 \quad 0 \quad 1 \quad 1 \quad 1 \quad 1 \quad 0 \quad 1 \quad 1 \quad 0.$$

The sequence of remainders is 1011110110. Checked against the sequence of powers of 2, this gives the representation of 758 as a sum of powers of two:

$$512 \quad 256 \quad 128 \quad 64 \quad 32 \quad 16 \quad 8 \quad 4 \quad 2 \quad 1$$
$$1 \quad 0 \quad 1 \quad 1 \quad 1 \quad 1 \quad 0 \quad 1 \quad 1 \quad 0$$

So, $758 = 512 + 128 + 64 + 32 + 16 + 4 + 2$.

2B: *Triangular numbers and large squares*

Represent the triangular number by a half-rectangle of dots whose sides differ by one unit. Two triangles make a rectangle and four

Fig. 2.6

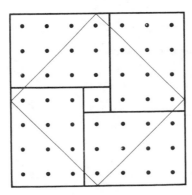

rectangles arranged round a central dot make a geometrical square. Therefore $8T + 1$ is always a perfect square.

2C: *Triangular numbers and cubes*

By observation, the differences between the squares of the triangular numbers are the cubes:

Triangulars (T)	1	3	6	10	15	21	$\cdots$
T^2	1	9	36	100	225	441	$\cdots$
Differences		8	27	64	125	216	$\cdots$

This can be proved by a bit of algebraic technique:

$$
\begin{aligned}
T_n^2 - T_{n-1}^2 &= [\tfrac{1}{2}n(n + 1)]^2 - [\tfrac{1}{2}(n - 1)n]^2 \\
&= \tfrac{1}{4}n^2[(n + 1)^2 - (n - 1)^2] \\
&= \tfrac{1}{4}n^2 \times 4n = n^3.
\end{aligned}
$$

2D: *Triangular numbers and squares*

The square of a triangular number is 'not far from' the product of the triangular numbers on either side of it.

Triangulars	1	3	6	10	15	21	28	$\cdots$
Squares	1	9	36	100	225	441		$\cdots$
$T_n^2 - T_{n+1}T_{n-1}$	1	3	6	10	15	21		$\cdots$

The initial 1 in the last sequence is calculated by supposing that the 0th triangular number is 0, which makes sense since it will contain no dots at all.

The differences reproduce the original sequence. This is a very common feature of such sequences. The result can be proved by algebra, as usual:

$$
\begin{aligned}
T_n^2 - T_{n+1}T_{n-1} &= [\tfrac{1}{2}n(n + 1)]^2 - \tfrac{1}{2}(n + 1)(n + 2) \times \tfrac{1}{2}(n - 1)n \\
&= \tfrac{1}{4}n(n + 1)[n(n + 1) - (n - 1)(n + 2)] \\
&= \tfrac{1}{4}n(n + 1) \times 2 = \tfrac{1}{2}n(n + 1).
\end{aligned}
$$

2E: *Sums of triangular sequences*

Calculation shows that

$$
T_1 + T_2 + T_3 = T_4
$$
$$
T_5 + T_6 + T_7 + T_8 = T_9 + T_{10}
$$
$$
T_{11} + T_{12} + T_{13} + T_{14} + T_{15} = T_{16} + T_{17} + T_{18}
$$
$$
T_{19} + T_{20} + T_{21} + T_{22} + T_{23} + T_{24} = T_{25} + T_{26} + T_{27} + T_{28}
$$
$$
T_{29} + T_{30} + T_{31} + T_{32} + T_{33} + T_{34} + T_{35} = T_{36} + T_{37} + T_{38} + T_{39} + T_{40}
$$
$$
\cdots
$$

The sequence 1, 5, 11, 19, 29, ... appears in the example following Problem 2E in the text. The sequence 2, 6, 12, 20, 30, ... is just double the triangular numbers.

2F: *An impossible sum-of-squares*

The square of an integer, such as 10, that is divisible by 2 but not by 4, will be divisible by 4 but not by 8, so it will leave a remainder 4 on division by 8. The square of an integer divisible by 4, such as 12, will be divisible by 16 and therefore also by 8, and so will leave no remainder. That deals with the even integers.

The square of an odd number, call it $2n + 1$, will be

$$(2n + 1)^2 = 4n^2 + 4n + 1 = 4n(n + 1) + 1.$$

Now one of the numbers n or $n + 1$ must be even, so their product is even, and $4n(n + 1)$ must be divisible by 8. Therefore the square of an odd number leaves remainder 1 when divided by 8.

Therefore, when three squares are added together, their total remainder when divided by 8 will be the sum, not exceeding 7, of three of the numbers 4, 0 and 1 (with one or two of these numbers quite possibly repeated). Therefore, the sum of three squares can have the remainders 0, 1, 1 + 1, 1 + 1 + 1, 4, 4 + 1, 4 + 1 + 1, only. It cannot have remainder 7. Therefore none of the integers in the sequence 7, 15, 23, 31, ... can be the sum of three squares.

2G: *Sums and differences of squares*

Take any odd square and write it as the sum of two consecutive integers, for example

$$7^2 = 49 = 25 + 24.$$

Then, since $25 - 24 = 1$, we have also

$$7^2 = (25 + 24)(25 - 24) = 25^2 - 24^2$$

and so

$$7^2 + 24^2 = 25^2.$$

This produces an infinite number of Pythagorean triangles, one for each odd square, but it does not produce all of them. In fact it produces only those which are approximately isosceles because their hypotenuse and one side differ by 1.

2H: *A formula for Pythagorean triples*

Any pair of consecutive odd or even numbers can be written as $n - 1$ and $n + 1$. The sum of their reciprocals will be

$$1/(n - 1) + 1/(n + 1) = (n + 1 + n - 1)/(n - 1)(n + 1)$$
$$= 2n/(n^2 - 1).$$

The numbers $2n$ and $n^2 - 1$ are sides of a right-angled triangle, because they are special cases of the formulae $2mn$ and $n^2 - m^2$, in which $m = 1$. As before, this method only produces some Pythagorean triangles, not all of them.

2I: *Sums of sequences of squares*

The pattern continues

$$3^2 + 4^2 = 5^2$$
$$10^2 + 11^2 + 12^2 = 13^2 + 14^2$$
$$21^2 + 22^2 + 23^2 + 24^2 = 25^2 + 26^2 + 27^2$$
$$36^2 + 37^2 + 38^2 + 39^2 + 40^2 = 41^2 + 42^2 + 43^2 + 44^2$$
$$\vdots$$

The sequence 3, 10, 21, 36, ... consists of every other triangular number, missing out 1, 6, 15, 28, ... The sequence of numbers immediately to the left of the equality signs is 4, 12, 24, 40, ... This is just four times the triangular numbers, and the sequence of pairs sandwiching the equality signs, 4–5, 12–13, 24–25 are the same pairs met in Problem G.

Taking the third line as an example, we get

$$21^2 = 25^2 - 24^2 + 26^2 - 23^2 + 27^2 - 22^2$$

or

$$21^2 = 49 \times 1 + 49 \times 3 + 49 \times 5$$

or

$$21^2 = 7^2(1 + 3 + 5) = 7^2 \times 9 = 7^2 \times 3^2.$$

In general, start with the difference (here $25^2 - 24^2$) of the squares of any pair whose sum is an odd square (here 24, 25), and consecutively add the ascending squares and subtract the descending squares ($+ 26^2$ and $- 23^2$, then $+ 27^2$ and $- 22^2$) until the total reaches the next descending square (21^2); this will always be possible because the sums of these pairs will always be the same (in this case 49) and their differences will be the sequence of odd numbers, 1, 3, 5, whose sums are always squares. The general pattern is

$$[n(2n + 1)]^2 + \cdots + [2n(n + 1)]^2 = (2n^2 + 2n + 1)^2 + \cdots + (2n^2 + 3n)^2.$$

Chapter 3

Mathematics as science

The easiest way to discover mathematical facts and theorems is to treat mathematical objects just as if they were objects in the real world, and to make observations and do experiments, by drawing and measuring on geometrical figures, and making calculations with numbers (or in algebra) just as we did in the first two chapters. On the basis of your experience, you make a *conjecture*, just like any scientist. You construct a little theory about 'what is the case', and then you either test it further, or, if you are sufficiently confident that it is true, you set out to prove it, by a persuasive argument. There is nothing unmathematical about all this, contrary to the impression anyone might get from studying traditional mathematics *textbooks*, in which everything is proved and nothing is ever guessed.

There is a difficulty, however, which has to be borne in mind. As long as you are behaving like a scientist, you can never be quite certain that your mathematical results are true. Therefore, scientific practice, although it is an essential tool of the mathematician, is also a source of danger. The mathematician who is fast asleep will not see the danger lurking, and will fall straight into the traps and snares that await. The mathematician who is wide awake, however, will gain the double benefit from scientific experiment and observation, of discovering all sorts of supposed facts, and gaining some idea *why* they might be true.

Mathematicians have always been ready to make observations, and to use *induction*; to put the same point another way, they have always made and tested *hypotheses*. This was made clear by James Sylvester (1814–97) in an address to the British Association for the Advancement of Science in 1869 made in response to the views of Thomas Huxley. Huxley was a great biologist, but he had no experience of actually doing mathematics. He thought that mathematics was nothing more than a matter of deduction from a few agreed principles, and that 'Mathematics is the study which knows nothing of observation, nothing of experiments, nothing of induction, nothing of causation.' Sylvester replied:

I think no statement could have been made more opposite to the undoubted facts of the case, that mathematical analysis is continually invoking the aid of new principles, new ideas and new methods, . . . that it is unceasingly calling forth the faculties of observation and comparison, that one of its principal weapons is induction, that it has frequent recourse to experimental trial and verification, and that it affords a boundless scope for the exercise of the highest efforts of imagination and invention.

There is a double historical irony here. T. H. Huxley was himself a great scientist who contributed especially to the study of what the anatomists called *homologous parts*: features which, although superficially different, actually displayed the same basic structure. Neither he nor Sylvester realized that the study of analogous structures was soon to become one of the major features of mathematics also. The second irony is that, in 1906, a few years after Sylvester's death, a manuscript was discovered in Constantinople containing a lost work of the great Greek mathematician Archimedes (287–212 BC) titled *The Method*, in which he describes a method of discovering certain mathematical theorems by mechanics, by treating geometrical shapes as if they were physical objects to be weighed and balanced against each other.

The very first of his discoveries by this method was the theorem that 'any segment of a parabola is four-thirds [in area] of the triangle which has the same base and equal height.' The theorem is illustrated on the left of Fig. 3.1, while the right-hand figure shows how Archimedes thought about it. He assumed that the parabolic segment is a thin plane sheet of material. A line AK through A parallel to the axis of the parabola meets the side CB produced at K, and CBK is produced to H, so that K is the midpoint of CH. In a wonderful feat of imagination, Archimedes considered HKC as the arm of a balance, swinging about its midpoint K, and asked himself what weight at H would balance the weight of the parabolic segment.

Fig. 3.1

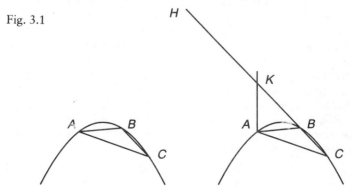

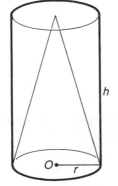

volume of cylinder = $\pi r^2 h$

volume of cone = $\frac{1}{3}\pi r^2 h$

Fig. 3.2

We shall not explain Archimedes' proof, but merely observe that he is behaving here like a scientist, making analogies, performing an experiment, and drawing scientific conclusions which he later proved by mathematical arguments. He emphasized that:

> This procedure is no less useful even for the proof of the theorems themselves; for certain things first became clear to me by a mechanical method ... it is of course easier, when we have previously acquired by the method some knowledge of the question, to supply the proof than it is to find it without any previous knowledge.

How very true! It is so much easier to solve a problem when you know in advance the answer you are trying to reach, which is why Archimedes split the credit for discovering that a cone is one third of the volume of the cylinder on the same base, and with the same height, between Democritus who discovered this fact and Eudoxus who proved it. (See Fig. 3.2.)

Sylvester added to his encomium of experiment in mathematics the statement that 'induction and analogy are the special characteristics of modern mathematics.' We are indeed very liable to spot patterns and analogies – because that is the way our brains work – and to fondly imagine that because they are so neat and pretty, that they must be correct. But this is far from always being the case!

As Henri Poincaré observed, 'When a sudden illumination invades the mathematician's mind ... it sometimes happens ... that it will not stand the test of verification ... it is to be observed almost always that this false idea, if it had been correct, would have flattered our natural instincts for mathematical elegance.' Aye, there's the rub! The mathematician expects sound mathematics to be elegant and pleasing, but the converse is not always true. A lot of attractive ideas are thoroughly unsound.

∗

Two hundred years before Poincaré, John Wallis, Savilian Professor of Geometry at Oxford University and an important predecessor of Newton, had recognized the attractions and the dangers of pattern-spotting. He wrote in his *A Treatise of Algebra* (1685):

I look upon *Induction* as a very good Method of *Investigation*; as that which does very often lead us on to the easy discovery of a General Rule; . . . And where the Result of such Inquiry affords to the view, an obvious discovery; it needs not (though it may be capable of it) any further Demonstration . . . where there is no ground of suspicion why it should fail . . .

Wallis gave as an example of an induction the very pattern of squares and cubes and their differences that we saw in the last chapter. Finding that the first difference of the squares increased steadily by 2,

---

Problem 3A *A pattern of powers*

The successive differences of the cubes can be written like this, to show a different pattern. Instead of simply writing $27 - 8 = 19$ and $64 - 27 = 37$, we leave them as they are and write the calculation in full as $(64 - 27) - (27 - 8) = 64 - 2 \times 27 + 8$:

cubes	1		8		27		64		125
1st diffs		$8 - 1$		$27 - 8$		$64 - 27$		$125 - 64$	
2nd diffs		$27 - 2 \times 8 + 1$		$64 - 2 \times 27 + 8$		$125 - 2 \times 64 + 27$			
3rd diffs			$64 - 3 \times 27 + 3 \times 8 - 1$			$125 - 3 \times 64 + 3 \times 27 - 1$			

Given that the third differences are all equal, what does this tell us about successive cubes? What is the connection with *Pascal's Triangle*, in which the ends of the rows are all 1s, and each other entry is the sum of the two numbers above it to the right and left in the previous row?

```
              1
            1   1
          1   2   1
        1   3   3   1
      1   4   6   4   1
    1   5  10  10   5   1
            . . .
```

and that the second differences of the cubes increase steadily by 6, he concludes:

> Such observations would be looked upon, as sufficiently instructive; there is no reason of Suspicion, why it should not so continually proceed; but reason rather to believe, that there is, in the nature of Number, a sufficient ground of such Sequel . . .

Was Wallis justified in his confidence, *in this case*? Yes, he was, in this case. But how could he be so certain? Wallis, as it happens, presented this argument in a diatribe against Pierre Fermat, who had his own problems with induction, though he did not realize it.

Fermat had claimed in a letter in 1640 to his friend Frénicle that $2^n + 1$ would have a factor if n was not a power of 2 (which is true) and also that, if n was a power of 2, then it would be a prime number, though Fermat admitted that he could not prove this last statement.

Taking n to be the powers of 2 in sequence, we easily calculate that:

$$2^1 + 1 = 3, \qquad 2^2 + 1 = 5,$$
$$2^4 + 1 = 17, \qquad 2^8 + 1 = 257, \qquad 2^{16} + 1 = 65{,}537.$$

These are indeed all prime. What a tempting pattern! If only it were true, there would be a simple connection between some primes and powers of 2, and we could easily calculate an infinite number of prime numbers – in theory! – and calculate some very large primes in practice, with complete confidence.

This would have been a remarkable achievement. The pattern of the prime numbers is subtly irregular and no one has ever found a simple formula which genuinely produces an infinite sequence of prime numbers. One of Fermat's difficulties was that the mathematicians had not yet learnt just how tricky the prime numbers are, and so Fermat could be as optimistic about his 'prime number sequence' as John Wallis had been about the differences of the cubes.

Another problem lay in the difficulty of factorizing large numbers. Although school children can do large multiplications, the opposite process of factorizing numbers is usually very difficult. Fermat's numbers become very large, very quickly. The very next Fermat number is

$$2^{32} + 1 = 4{,}294{,}967{,}297$$

and the next, $2^{64} + 1$, is a number of 20 digits. Numbers this large, and much larger, can be factorized easily nowadays using computers. For example,

$10^{20} + 1 = 100000000000000000001$
$= 73 \times 137 \times 1676321 \times 5964848081,$

$10^{60} + 1 = 73 \times 137 \times 1676321 \times 99990001 \times 5964848081$
$\times 10000999999989999899990000000010001.$

There is some pattern here, because the factors of $10^{20} + 1$ are all factors of $10^{60} + 1$, but that is all. How could Fermat have been so confident that such large numbers were prime?

Fermat made many conjectures, and they were almost always correct, but not in this case. Fermat had made one of the most notorious mistakes in the entire history of mathematics.

Leonhard Euler showed in 1732 that $2^{32} + 1 = 4,294,967,297 = 641 \times 6,700,417$. Nearly one hundred and fifty years later, in 1880, F. Landry showed that $2^{64} + 1$ is the product of two primes: $274,177 \times 67,280,421,310,721$. The next Fermat number, $2^{128} + 1$ has 39 digits: using an electronic computer, its factors were only found as recently as 1970.

The next dozen Fermat numbers are known to be composite, although most of their actual factors are unknown. It is quite plausible that all the remaining Fermat numbers are composite, so that only those up to $2^{16} + 1$ would be prime – just five!

Mathematicians have good reason to be suspicious of results obtained by scientific methods. They can never know, by science alone, that they are certainly true. At best they can rely on their *intuition*, or their sense of what is elegant and 'sure to be correct', but as Poincaré noted, they are often led astray by such feelings.

Here is an example: 1/7 as a decimal is 0.142857 142857 142857 . . . The period, 142857, repeats endlessly. If you had any reason to calculate 17^5 on a calculator, you will find that it is 1419857. Well, really! Can it be a coincidence that these two numbers are so similar?

Another case: every prime number is one more or one less than a multiple of 6, because numbers of the forms $6n$, $6n + 2$, $6n + 3$ and $6n + 4$, are at least divisible by 6, 2, 3 and 2, respectively.

Problem 3B   *Fermat numbers*

There is another relationship which can be used to define the sequence of Fermat numbers. Without knowing, or observing, any connection with powers of 2, how can 5 be calculated from 3; 17 from 5 and 3; and so on?

For small multiples of 6, a sort of converse also seems to be true. Take any multiple of 6, say 48, and either 47 or 49, or both, will be a prime. In this case, 47 is prime.

This works all the way up to 100 and beyond. How suspicious should we be of this pattern? Very suspicious, as it happens. Because the prime numbers get scarcer and scarcer the higher we go, while the multiples of 6 do not, we will eventually find that there are not enough primes to match all the multiples of 6. In fact, 120 is the first counterexample: $119 = 7 \times 17$ and $121 = 11^2$.

Our next example is called the 'prime number race'. Write down the prime numbers in sequence in two rows, primes that are one less than a multiple of 4 in the top row, primes that are one more in the bottom row:

```
3  7 11    19 23 31    43 47  59  67 71  79 83         103 107      127 · · ·
5     13 17     37 41      53  61     73     89 97 101      109 113· · ·
```

Counting each row, it seems that there are more of the primes $4n - 1$ in the top row; they are always winning the race. Checking with a computer or in a table or primes as far as 25,000 and beyond suggests that this is always the case. Not so! On reaching 26,849 their numbers are equal, and then 26,861 puts the $4n + 1$ team in the lead, but only momentarily, because the next two primes are both $4n - 1$.

The first region within which the $4n + 1$ are consistently in the lead stretches from 616,769 to 633,881. In fact the lead in this race changes infinitely often, as Littlewood proved in 1914.

Bearing in mind these facts, would you expect a typical odd number, picked at random, to have more divisors of the form $4n - 1$ or $4n + 1$? We are now talking of *divisors*, so 9 has three divisors, 1, 3 and 9, of which 3 is a $4n - 1$, and 1 and 9 are $4n + 1$. The number 15 has divisors 1, 3, 5, 15 which is two of each, while 21 has 1, 3, 7, 21, also two of each; 25 has divisors 1, 5, 25 which are all $4n + 1$, and 17 has 1 and 17 which is two of $4n + 1$, only.

The answer is a surprise. Legendre (1752–1833) proved that, if the number of $4n + 1$ divisors is D and the number of $4n - 1$ divisors is d, then the number can be expressed as the sum of two squares in $4(D - d)$ ways. In this formula, different orders of the numbers and different signs (positive and negative) are all counted separately, so for our last example (17), $D = 2$, $d = 0$, $4(D - d) = 8$ and $17 = (+4)^2 + (+1)^2$ plus the seven variants on this obtained by exchanging the 4 and 1 and changing one or both of their signs from + to −.

This formula tells us what we want to know, because the number of ways of expressing an integer as the sum of two squares is always either zero or greater than zero. It says that $4(D - d) \geqslant 0$, so $D - d \geqslant 0$ and D is always greater than or equal to d!

The factorials are calculated by multiplying the integers together, in sequence. Thus 1 factorial, written 1! (and often read as one-bang!) is just 1. The first few factorials are listed in Table 3.1.

Table 3.1

1!	= 1
2! = 2 × 1	= 2
3! = 3 × 2 × 1	= 6
4! = 4 × 3 × 2 × 1	= 24
5! = 5 × 4 × 3 × 2 × 1	= 120
⋮	
10! = 10 × 9 × 8 × 7 × 6 × 5 × 4 × 3 × 2 × 1	= 3,628,800

Problem 3C *A geometrical picture*

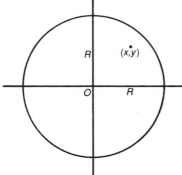

Fig. 3.3

Most integers are not the sum of two squares and therefore, by Legendre's formula, they have the same number of $4n + 1$ divisors as they have $4n - 1$ divisors. What proportion of numbers *are* the sum of two squares? We can estimate this by representing each number $x^2 + y^2$ by the point (x, y) on the graph shown in Fig. 3.3. By Pythagoras' theorem, if $x^2 + y^2$ is less than R^2, then the point (x, y) will lie inside a circle whose centre is $(0, 0)$ and whose radius is R.

How can we roughly estimate, from the area of this circle, the proportion of integers less than R^2 that are the sum of two squares?

The following calculations suggest that we might be able to calculate new primes indefinitely from the factorials:

$$3! - 2! + 1! = 5$$
$$4! - 3! + 2! - 1! = 19$$
$$5! - 4! + 3! - 2! + 1! = 101$$
$$6! - 5! + 4! - 3! + 2! - 1! = 619$$
$$7! - 6! + 5! - 4! + 3! - 2! + 1! = 4421$$
$$8! - 7! + 6! - 5! + 4! - 3! + 2! - 1! = 35,899.$$

The numbers 5 to 35,899 are all prime. Is this pattern going to continue? Is it genuine? The case of Fermat's 'prime' number sequence should be adequate warning. The very next sum is

$$9! - 8! + 7! - 6! + 5! - 4! + 3! - 2! + 1! = 326,981$$
$$= 79 \times 4139.$$

The next example is geometrical. Figure 3.4 shows 1, 2, 3, 4 and 5 dots marked round separate circles, and joined to form regions. The first four circles are divided into 1, 2, 4 and 8 regions respectively. In the fifth circle, as careful counting will show, there are (rather naturally?!) 16 regions. How many regions will there be in the sixth circle? What is the general rule? This pattern seems very simple, and very strong, and it has nothing to do with prime numbers. Yet it breaks down immediately. The very next case, with 6 dots on the circle, contains only 31 regions (Fig. 3.5). The general formula for the number of regions is:

$$\tfrac{1}{24}(n^4 - 6n^3 + 23n^2 - 18n + 24).$$

Fig. 3.4

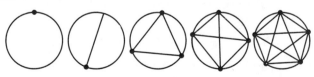

Fig. 3.5

The sequence continues

$$1 \quad 2 \quad 4 \quad 8 \quad 16 \quad 31 \quad 57 \quad 99 \quad 163 \quad 256 \quad 386 \quad 562 \quad \cdots$$

It was Euler who originally discovered that the quadratic formula $n^2 + n + 41$ has a large number of prime values. In fact the values are prime for any value of n from -40 to $+39$. What happens then? Well, $40^2 + 40 + 41 = 41 \times 40 + 41$, which is not prime, and likewise $41^2 + 41 + 41$ is obviously divisible by 41, so the pattern breaks down, but not completely: many values of the formula for larger values of n are also prime:

n	38	39	40	41	42	43	44	45	$\cdots$
$n^2 + n + 41$	1523	1601	*	*	1847	1933	*	2111	$\cdots$

When a pattern which seems strong breaks down, it is natural to wonder why. Sometimes the breakdown is not so very surprising. There are many squences, for example, which start with a few powers of 2, and then diverge. For example, in the sequence

$$1 \quad 2 \quad 4 \quad 7 \quad 11 \quad 16 \quad 22 \quad \cdots$$

the differences between the terms are increasing by 1, each time. It starts 1–2–4 and then diverges. In the sequence

$$1 \quad 2 \quad 4 \quad 8 \quad 15 \quad 28 \quad 42 \quad \cdots$$

each term is one more than the sum of the previous three terms. There are many sequences with this rule, because we can choose any numbers we like for the first three numbers. This sequence just happens to start 1–2–4.

The sequence $n^2 + n + 41$ is much more remarkable, and we can expect a remarkable explanation of its near-success. There is one, though it is far too complicated to more than indicate here:

$$e^{\pi\sqrt{163}} = 262{,}537{,}412{,}640{,}768{,}743.999{,}999{,}999{,}999{,}250 \ldots$$
$$163 = 4 \times 41 - 1.$$

At this point there should be a fanfare of trumpets for one of the most profound connections in all of mathematics! In the first line, e is the base of natural logarithms, which can be found on most calculators by entering 1 and then pressing the key marked e^x. The value of e is approximately 2.718281828 . . . (no, *that* pattern in its digits does not continue). Powers of e are not usually close to being an integer. Charles Hermite (1822–1901) showed that $e^{\pi\sqrt{163}}$ is indeed close to being an integer while he was working on what is known as the class

Problem 3D *Bundles of arrows*

The Indian mathematician Mahavira, who flourished about AD 850, posed this puzzle about the number of arrows in a bundle in his book *Ganita-Sara-Sangraha.*

The circumferential arrows are eighteen in number. How many in all are the arrows to be found in the bundle?

Such problems had more than a mere curiosity value to soldiers. As late as the eighteenth century in Europe, manuals of mathematics showed how the number of cannonballs in a pile could be calculated from the length of one side, depending on whether they were piled in a triangular (Fig. 3.6) or a square pyramid. Armed with these formulae, the spy with a telescope could estimate the amount of ammunition the enemy had available.

Fig. 3.6

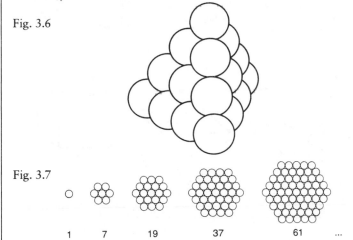

Fig. 3.7

| 1 | 7 | 19 | 37 | 61 | ... |

Leaving aside these military uses of mathematics, the answer to Mahavira's puzzle is illustrated in Fig. 3.7. After solving the puzzle, look again at the sequence. You might notice that:

$$1 + 7 =\ \ 8 = 2^3$$
$$1 + 7 + 19 = 27 = 3^3$$
$$1 + 7 + 19 + 37 = 64 = 4^3$$
$$\vdots$$

The problem is to decide whether this pattern is genuine, or another false trail.

number problem. It is typical and entirely natural that, since that problem is so very difficult, its solution involved the discovery of a great deal of remarkable mathematics.

The next case also proved difficult, and led to remarkable developments. The *polyhedra* shown in Fig. 3.8 have different numbers of faces, edges and vertices. However, if we make a table of these (Table 3.2), then we can rather easily spot a pattern.

Fig. 3.8

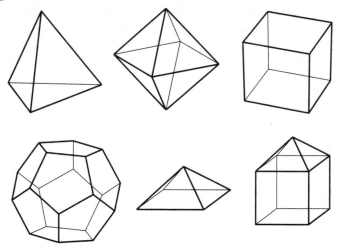

Table 3.2

	Faces	Vertices	Edges	Faces + vertices − edges
Tetrahedron	4	4	6	2
Octahedron	8	6	12	2
Cube	6	8	12	2
Dodecahedron	12	20	30	2
Square pyramid	5	5	8	2
Cube + pyramid	9	9	16	2

This pattern is $F + V = E + 2$. It is known as Euler's relationship – yet another result named after the great man. From this pattern – if it is sound – we can make many deductions, so it is potentially very important. For example we can deduce that there are only five regular polyhedra (a polyhedron is regular if all its faces and all its vertices are identical). Another consequence is that, in a very large but finite plane map in which at most three countries meet at any point, the average number of countries neighbouring a country will be slightly under 6.

Suppose that on such a very large map there are T triangles, Q quadrilaterals, P pentagons, H hexagons, and so on. Then

$$\text{number of faces} = T + Q + P + H + \cdots$$
$$\text{number of edges} = \tfrac{1}{2}(3T + 4Q + 5P + 6H + \cdots)$$
$$\text{number of vertices} = \tfrac{1}{3}(3T + 4Q + 5P + 6H + \cdots).$$

The fractions $\frac{1}{2}$ and $\frac{1}{3}$ appear because every edge is shared between two polygons and every vertex is shared between three. We ignore the boundaries of the map because it is very large, and so the boundary polygons form a very small proportion of the total. By Euler's relationship,

$$(T + Q + P + \cdots) + \tfrac{1}{3}(3T + 4Q + 5P + \cdots)$$
$$= 2 + \tfrac{1}{2}(3T + 4Q + 5P + \cdots).$$

So, $(T + Q + P + \cdots) = 2 + \tfrac{1}{6}(3T + 4Q + 5P + \cdots)$ and hence

$$6(T + Q + P + \cdots) = 12 + (3T + 4Q + 5P + \cdots).$$

The average number of edges per polygon will the total of edges (counted separately for each polygon) divided by the total number of faces. That is,

$$(3T + 4Q + 5P + \cdots)/(T + Q + P + \cdots)$$
$$= 6 - 12/(T + P + Q + \cdots).$$

Problem 3E   *Faces and vertices*

Three of the regular polyhedra have three faces meeting at each vertex, the smallest possible number. How can this fact be deduced from Euler's relationship, by supposing that the regular polyhedron has F faces and deducing the number of edges and vertices?

In other words, it is slightly less than 6. This result has been checked against the municipal counties of Brazil, of which there are 2800. In a random sample of these, excluding counties which were on the the sea, the number of neighbours varied from two to fourteen, and nearly one third of them had exactly 6 neighbours. The average number of neighbours was 5.71.

*

Here is another deduction, which uses almost exactly the same argument as the last. Suppose that you are designing a giant geodesic dome, consisting of a framework of H hexagons and P pentagons, with three faces meeting at each vertex. You wonder, naturally, how many of each shape you will require.

The total number of faces will be $H + P$. Each hexagon has six edges and each pentagon five, but each edge is shared between two faces, so the total number of edges will be $\frac{1}{2}(6H + 5P)$. Similarly, the total number of vertices will be $\frac{1}{3}(6H + 5P)$. From Euler's relationship,

$$H + P + \tfrac{1}{3}(6H + 5P) = \tfrac{1}{2}(6H + 5P) + 2$$

Simplifying, we discover that $P = 12$. In other words, however many hexagons we use in total, and it could be hundreds, there must be exactly 12 pentagons. We are not told where they must go – perhaps

---

Problem 3F *Rigid frameworks*

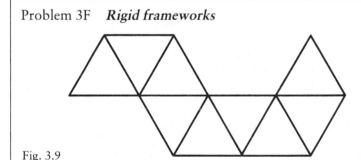

Fig. 3.9

We saw in Chapter 1 that, if a plane triangular framework is just rigid, and it has n joints, they will be joined by $2n - 3$ bars. In Fig. 3.9, $n = 11$ and there are 19 bars. How many triangles will be formed by the framework? What will be the average number of bars meeting at each vertex? How can the answer be reconciled with a tessellation of equilateral triangles with an average of nearly 6 edges per vertex?

they can go anywhere? – but there must be exactly 12 in the complete geodesic dome.

We have already seen a special case of such a dome: it is the regular dodecahedron. It is certainly true that all its faces are either hexagons or pentagons. It just happens that the number of hexagons is zero, and there are twelve pentagons. These happen to be regular, but we could easily change that by distorting the dodecahedron. The faces would then be irregular, but Euler's relationship and our conclusion would still hold.

Euler's relationship is clearly important, if it always works. The question remains: how convincing is it? How suspicious should we be? Cauchy in 1813 saw no reason to be suspicious and published a 'proof' which didn't work. He had failed to consider the possibility that the polyhedron might have a hole, as in Fig. 3.10. This hollow polyhedron is a square slab with a hollow cube at its centre. The top, bottom and both sides are plane, and the front and back faces are identical. It has 16 vertices, 16 faces and 32 edges, so $V + F = E$. This looks like Euler's relationship, but it is not quite right. Naturally we wonder whether other hollow shapes give the same result. The one in Fig. 3.11 does.

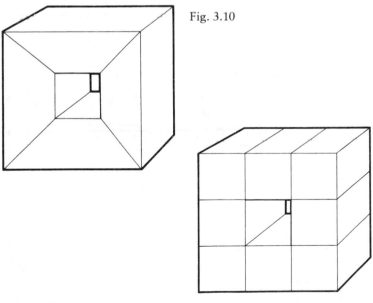

Fig. 3.10

Fig. 3.11 $V = 32, F = 32, E = 64$

This is similar to the first, but it is made up by fixing eight cuboids (rectangular bricks) face to face. It has 32 vertices and 32 faces and 64 edges, so the 'modified formula' again works.

What Cauchy overlooked, S. A. J. Lhuilier realized, at almost exactly the same time. In fact he noticed three exceptional cases: the ring-shaped polyhedra and two others. He had been observing mineral crystals. Crystals are found in polyhedral forms, so it was likely that they would occasionally exhibit unusual and exceptional shapes. Lhuilier found that occasionally an opaque smaller crystal was surrounded by a translucent larger crystal. The resulting polyhedron of the outer crystal was certainly a solid bounded by polygonal faces (Fig. 3.12), so it should surely fit Euler's relationship. But it doesn't. This cube-in-a-cube has twice as many edges, vertices and faces as a single cube, so $V + F - E$ will be twice as large. That is $V + F = E + 4$. Crystals also occasionally have indentations in their faces. The 'polyhedron' in Fig. 3.13 has a pyramidal indent, and $V + F = E + 3$. (The same would apply if the pyramid pointed outwards.) Were these the only counterexamples? Not at all. Twenty years later

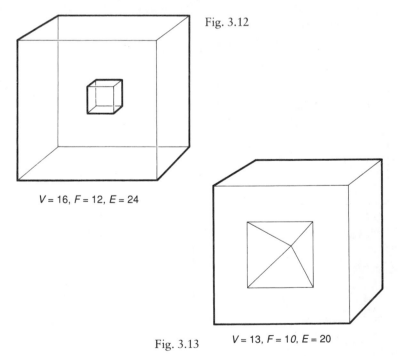

Fig. 3.12

$V = 16, F = 12, E = 24$

$V = 13, F = 10, E = 20$

Fig. 3.13

Problem 3G *Buckminsterfullerene and a familiar shape*

How can twelve regular pentagons, and a number of hexagons be arranged to form a semi-regular solid? 'Semi-regular' means that, although the faces are not all identical, all the vertices are the same. In this case, the same number of hexagons will surround each pentagon, and the same number of pentagons will be adjacent to each hexagon.

The sought-for solid has the shape of a hollow carbon molecule discovered in 1985 by a group of chemists who named it Buckminsterfullerene in honour of the inventor of the geodesic dome. It is also the shape of several types of viruses, and of a familiar object owned by many teenagers, especially boys. What is it?

Hessel considered examples such as that in Fig. 3.14. In this case, some of the faces need to be twisted to match up with the edges; however, curvature of the faces does not in itself affect the validity of the Euler-type formulae. On the left is an ordinary box, with the long sides divided into two faces each. If we squeeze the centre of the box, the four central edges will eventually become a straight line as in the right-hand figure. The solid box is still a solid bounded by (twisted) polygons, but it has two insides, as it were. Four central edges have become two edges, but the four central vertices have become three vertices, and now $V + F = E + 3$.

Fig. 3.14

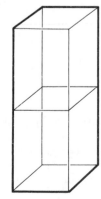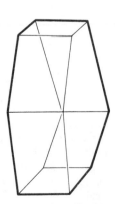

$V = 11, F = 10, E = 18$

Fig. 3.15

These examples are not the end of the matter. Two hundred years earlier, Kepler had described in his *Harmonice Mundi*, more famous for containing his laws of planetary motion, the two polyhedra shown in Fig. 3.15: the small stellated dodecahedron and the great stellated dodecahedron.

Problem 3H *Mountains, valleys and ridges*

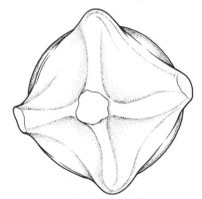

Fig. 3.16

Figure 3.16 shows one of the minor planets which orbit between Mars and Jupiter. On the surface can be seen several mountain peaks, several valleys and the mountain ridges that separate one valley from the next. If there are 10 mountain peaks, and 12 valleys on the surface, can you say how many ridges there are, without seeing the entire surface?

Fig. 3.17

Kepler, and mathematicians since, had seen these as having pentagonal faces in the form of star pentagons. Each regular star pentagon has five sides and five equal angles, so it fits the obvious definition of a regular pentagon, although it is different from the usual regular pentagon and does cross itself (Fig. 3.17). The twelve faces of the stellated dodecahedra intersect themselves also. Does this mean they do not fit Euler's relationship? Indeed, they do not. The small stellated dodecahedron has 12 star-pentagonal faces, 12 vertices and 30 edges, each cut at two points by other faces. So $V + F - E = -6$.

Do these examples seem unfair? It is certainly going to extremes to make a polyhedron cut itself. But it is also going to extremes to almost separate a polyhedron into two separate parts, and then treat it as one, or to poke a hole right through a polyhedron and claim that the result is still a polyhedron. The fact is that the definitions of Euler's day and earlier did not recognize any of these variations. It seems that Euler's relationship works for a polyhedron without any holes, which does not intersect itself, and which has, as it were, only one 'inside'.

What do we do now? A polyhedron is usually defined as a closed shape enclosed by faces all of which are plane polygons. If we stick to that definition, then Euler's relationship is only true for *some* polyhedra. Alternatively, we could admit that there are polyhedra and polyhedra, and invent a new name for those that do not fit Euler's relationship. The ones that fit could be called 'normal polyhedra', and the others 'special' or 'abnormal'.

The point is that, by attempting to answer this problem, we have gone beyond the original issue, to question the very nature of polyhedral shapes, and their properties. Solving this question forced mathematicians to examine their ideas more closely, and they ended up having a much clearer idea of what a polyhedron 'is'.

The truth is that, when early mathematicians talked about polyhedra, they thought they knew what they were talking about, but they didn't. They only had a rather vague idea: that a polyhedron was an object like a cube, or tetrahedron, or pyramid.

This is very natural. In everyday life, most definitions are of this form. A sparrow is any bird that looks and behaves sufficiently like a sparrow. It takes an ornithologist to distinguish subtle variations, and maybe to decide whether a bird that does not seem to fit the usual 'picture' is indeed a sparrow, or another species entirely, and an ornithologist can only do this because scientists have previously studied different species in great detail, so that they understand

---

Problem 31   *A polyhedral cheese*

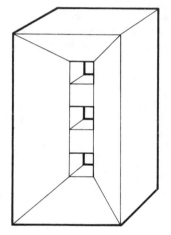 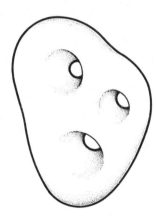

Fig. 3.18

Since Euler's relationship works after a fashion for all sorts of polyhedra, it seems natural to accept it as true, *provided that* the value of $V + F - E$ is adjusted for the type of polyhedron you are considering. For 'ordinary' polyhedra, which are equivalent to a sphere in a natural sense, $V + F - E = 2$. If there is just one hole going right through, and no internal hole, then, apparently, however the surface is divided up into polygons, $V + F - E = 0$. What will be the value of $V + F - E$ for polyhedra with more holes? The surface on the left of Fig. 3.18 is already constructed from polygonal faces. The surface on the right could be turned into a polyhedron by plastering its surface with polygons.

them much more deeply than the lay person. Mathematicians have the same problem when they meet mathematical objects or situations for the first time. Only deep study will reveal what is 'really the case'.

This process, of answering deep questions and so discovering what you really meant by what you've been talking about all the time, is very common in mathematics. It is also very important and very scientific. It has to be scientific, because you cannot prove that one definition is superior to another by logic: you can only make judgements that this definition is preferable to that. You could say that, until mathematicians have examined very deeply and scientifically the objects they talk about so glibly, then they do not really know what they are talking about.

After such a scientific examination, they know and can say exactly what they mean, and they are in a position to prove results without tying themselves in knots, although, of course – it has to be admitted – you can never be absolutely certain that your scientific examination of a mathematical object, or idea, is finally over! It could just be that, in years to come, a further examination will reveal important features that you had overlooked.

This happened at the turn of the century to the system of geometry presented by the Greek Euclid in his *Elements*. For more than two thousand years, the *Elements* had been regarded as a perfect example of mathematical reasoning, building up a structure of conclusions, on the basis of certain deductions from a few axioms. Early in the nineteenth century, several mathematicians discovered that Euclid's *parallel axiom* could not be proved, and that there were other kinds of geometry for which it did not hold, known collectively as non-Euclidean geometry. However, this was hardly a criticism of Euclid, because he had made quite clear that he was only assuming his parallel axiom, and not claiming to demonstrate it.

Then in 1899 Hilbert pointed out that, although Euclid assumed the idea of 'betweenness', he never actually said anything about it, so Hilbert filled in the gaps in Euclid's arguments by closely analysing what 'betweenness' means.

As it happens, Hilbert's improved analysis of Euclidean geometry did not undermine any of Euclid's results. Provided you make the assumptions that Euclid made, wittingly and unwittingly, his conclusions are all correct.

A few years ago Michael Atiyah, a great contemporary British mathematician whose *Collected Works* have just been published in six substantial volumes, and the American Saunders MacLane, were discussing how to do mathematics. MacLane set out what he thought was a standard position. You decide what area of mathematics you are interested in, you decide exactly what the terms that you are using mean, you know exactly what objects you are dealing with, and off you go.

Atiyah disagreed. As MacLane recorded,

Atiyah much preferred the style of the theoretical physicists. For them, when a new idea comes up, one does not pause to define it, because to do so would be a damaging constraint. Instead they talk around and about the idea, develop its various connections, and finally come up out with a much more supple and richer notion.

MacLane persisted: indeed, he seemed rather upset. But they had no need to disagree. Rather, Atiyah was talking about mathematics at the stage when you do indeed not know absolutely clearly what you are talking about. It is then wise to behave like a physicist and talk round it, walk round it, examining it closely.

Different mathematicians will also have different tastes and preferences which will drive them towards different kinds of mathematical research. As Atiyah said in an interview, 'Some people are very good at [using technical tools]; I'm not really. My expertise is to skirt the problem, to go around the problem, behind the problem . . . and so the problem disappears.' Imagine how much more quickly those early mathematicians would have resolved the problem of Euler's relationship if only they had realized that they had no clear idea of what a polyhedron is, and had instead taken Atiyah's approach!

Euclid never realized the weakness in his magnificent *Elements* that Hilbert pointed out. The early Greeks did discover, however, that there was a weakness in their idea of number. Early Greek mathematicians thought they knew very well what a number was. Numbers were either integers, formed in a sequence starting from 1, or they were ratios of integers – what we call fractions.

They also confidently 'knew' that every line has a length, measured by a number. So, for example, in the square of side 1, shown in Fig. 3.19, the length of the diagonal had to be measured by a number, which would be a fraction. It was a great disappointment to Pythagoras, therefore, but also one of the great triumphs of early mathematics, when he or one of his circle discovered that this was impossible. By Pythagoras' theorem,

$$AB^2 + BC^2 = CA^2$$

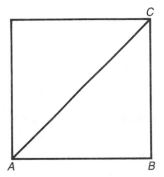

Fig. 3.19

That is, $CA^2 = 1^2 + 1^2 = 2$, and so

$$CA = \sqrt{2}.$$

The discovery that the square root of two, which is obviously not a whole number, could not be a fraction either was so disturbing that, according to legend, the secret was kept within the circle of Pythagoras and his disciples; when one student (Hippasus) who had let the secret out was drowned at sea, his fate was considered to be his deserved punishment.

Yet, although Pythagoras had failed in one direction, he (and/or his circle) had made a marvellous discovery – that of *irrational* numbers. His pupil Theaetetus later proved that the square roots of all the numbers up to 17, excluding 4, 9 and 16, are irrational. We know that most numbers are irrational. By failing, Pythagoras deepened our understanding of what numbers 'are'. His success had immeasurable consequences.

(We will see why all integers, other than perfect squares, have irrational square roots, in Chapter 11.)

Naturally, today we understand very clearly the idea of a number, and the idea of a length. Don't we? In 1954, Hugo Steinhaus, the author of the popular *Mathematical Snapshots*, wrote a paper with the deceptively simple title 'Length, shape and area', in which he imagined trying to measure *exactly* the length of the river Vistula. He argued that, 'the left bank of the Vistula, when measured with increased precision, would furnish lengths ten, hundred, or even a thousand times as great as the length read off the school map.' What Steinhaus had in mind is illustrated in Fig. 3.20. We can measure the coastline on the left, by marking round it with a pair of dividers

Fig. 3.20

opened to a fixed distance, and counting the number of steps. This is equivalent to approximating to the coastline by a polygon with many small equal sides.

Problem 3J *Is this a number?*

The decimals for fractions such as $\frac{1}{7}$, $\sqrt{2}$, or π are defined *from the left*. We know how they start, and we can calculate how they continue to the right:

$$\tfrac{1}{7} = 0.142857\,142857\,142857 \cdots ,$$
$$\sqrt{2} = 1.141213562373095 \cdots ,$$
$$\pi = 3.14159\,26535\,89793 \cdots .$$

Here is a method of defining a 'number' starting from the right-hand end.

$$\mathit{5}$$
$$2\mathit{5}$$
$$6\mathit{25}$$
$$39\mathit{0625}$$
$$152587\mathit{90625}$$
$$\cdots .$$

The rule is that each line is the square of the previous line. So this table tells us that it ends ... 90625. The digits in bold italic form a 'number' of which we only know the last few digits. Yet we can calculate more if we choose; in principle, the digit in each given position from the end is well defined. Does this process define a 'number'?

Here are three tests which will tell us more about this 'number' and help us to decide whether it should qualify as a proper number. Call the number D:

        A: calculate D^2

        B: calculate $E = 1 - D$, and calculate E^2

        C: calculate $D \times E$.

Now go round the same coastline, as in the second figure, with the length of each step reduced somewhat. The number of steps has considerably increased and the total length has also increased. Next, repeat with even smaller steps, blowing up the picture of the coastline if necessary to see exactly what you are doing . . .

This possibility has political importance. L. F. Richardson noted that pairs of neighbouring countries often claimed different lengths for their common borders: Spain had claimed 987 km for its border with Portugal, but the Portuguese claimed that the 'same' border was 1214 km long.

These cases are on a large scale; but everyone knows that, if you use a microscope, then even the side of a very smooth ruler (for example) will appear jagged and uneven. This suggests that, in the real world, no lengths of actual lines can be measured accurately without some reference to the scale of your measurements. Mathematical lines, such as the circumference of a circle, which can be measured accurately, are the exception, not the rule.

Yet, even in mathematics, lines exist which are far from smooth and easily measurable. Koch's snowflake curve, published in 1914, is an example (see Fig. 3.21). Start with an equilateral triangle, and divide each side into thirds, then construct equilateral triangles on each of the middle thirds. Repeat, by dividing all the new sides into thirds, and putting equilateral triangles on the middle thirds. Repeat, an infinite number of times.

At each step, every side is replaced by four smaller sides, whose total length is 4/3 of the original side, so the total length is multiplied repeatedly by 4/3. The total area of the final shape, however, if the original triangle has area 1, is only 8/5. After an infinite number of steps, the resulting snowflake curve has a finite area, but infinite length.

Is there anything more for mathematicians to learn about the simple idea of length? It is a notable fact, which has only been widely

Fig. 3.21

Problem 3K   *The reverse Koch snowflake*

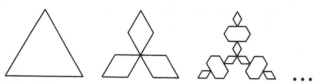

Fig. 3.22

The same process which generates the Koch snowflake curve can be performed on the inside of the triangle, instead of the outside. The resulting curve, after an infinite number of steps, will again be of infinite length. It also has an infinite number of *double points* where two small triangles which have been removed, share a vertex. The area remaining is obviously finite. How large is it?

realized in recent years, despite the pioneering insights of Sylvester and others, that mathematicians do behave remarkably like scientists when they are exploring their mathematical worlds, and observing and collecting and classifying, and doing experiments, and studying striking and even bizarre phenomena. Yet mathematics is *not* a science, or, at least, it is not *merely* a science. Mathematics has another aspect, a vital aspect, which links it to games such as chess and go, as we shall now see.

Solutions to problems, Chapter 3

3A: *A pattern of powers*

Take any sequence of numbers, and calculate successive rows of differences, and the numbers in Pascal's triangle appear:

$$
\begin{array}{cccccc}
a & b & c & d & e & f \\
b-a & c-b & d-c & e-d & f-e & \\
c-2b+a & d-2c+b & e-2d+c & f-2e+d & & \\
d-3c+3b-a & e-3d+3c-d & f-3e+3d-c & & & \\
e-4d+6c-4b+a & f-4e+6d-4c+b & & & & \\
f-5e+10d-10c+5b-a. & & & & &
\end{array}
$$

If there were more numbers in the top row, then correspondingly more rows of differences could be calculated.

In the present case, the numbers in the third row are all constant, and equal to 6; so, for any value of n,

$$n^3 - 3(n - 1)^3 + 3(n - 2)^3 - (n - 3)^3 = 6.$$

Since all rows after the third row are zero, the fifth row of differences, for example, which starts with,

$$6^3 - 5 \times 5^3 + 10 \times 4^3 - 10 \times 3^3 + 5 \times 2^3 - 1 \times 1^3$$

will equal zero, as can be checked by calculator.

3B: *Fermat numbers*

Each Fermat number is 2 more than the product of all the previous Fermat numbers:

$$3 + 2 = 5$$
$$3 \times 5 + 2 = 17$$
$$3 \times 5 \times 17 + 2 = 257$$
$$3 \times 5 \times 17 \times 257 + 2 = 65,537.$$

From the definition of Fermat numbers, the product of the first two is

$$(2 + 1)(2^2 + 1) = 2^3 + 2^2 + 2 + 1,$$

and so

$$(2 + 1)(2^2 + 1) + 2 = 2^3 + 2^2 + 2 + 1 + 1 + 1$$
$$= 2^3 + 2^2 + 2 + 2 + 1$$
$$= 2^3 + 2^2 + 2^2 + 1$$
$$= 2^3 + 2^3 + 1$$
$$= 2^4 + 1.$$

The same pattern works for the larger products also.

3C: *A geometrical picture*

We can choose to ignore negative values of x and y (unlike Legendre) so that we count, for example,

$$13 = 2^2 + 3^2 \quad \text{and} \quad 13 = (-2)^2 + 3^2$$

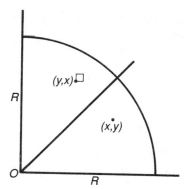

Fig. 3.23

as essentially the same. Then we are only interested in points (x,y) in the first quadrant, as in Fig. 3.23. Each such point is the lower-left corner of a unit square. So the number of integral points (x,y) in this quarter circle is approximately $\frac{1}{4}\pi R^2$, with a small adjustment for any points lying on the boundary of the quadrant.

However, we have still counted the points (x,y) and (y,x) as different, though we probably think of $x^2 + y^2$ and $y^2 + x^2$ as being the same. So we should halve this result, and the number of $x^2 + y^2$ less than R^2 is approximately $\frac{1}{8}\pi R^2$. For $R^2 = 100$, this comes to approximately 39.3, and a check on the table in Chapter 2 of sums of squares reveals that there are exactly 39 less than 100 (including pairs in which one square was zero, so that the point will lie on the boundary of the quadrant).

3D: *Bundles of arrows*

Mahavira's puzzle is answered by the fourth picture in Fig. 3.7. The total number of arrows in that bundle is $1 + 6 + 12 + 18 = 37$.

The nth number in the sequence is

$$1 + 1\times6 + 2\times6 + 3\times6 + \ldots + (n - 1)\times6$$
$$= 1 + 6[1 + 2 + 3 + \ldots + (n - 1)]$$
$$= 1 + 6T_{n1}.$$

This relationship can be illustrated by dividing each hexagonal number as in Fig. 3.24. It can also be used to find the formula for the nth hexagonal number:

$$1 + 6T_{n1} = 1 + 6 \times \tfrac{1}{2}(n - 1)n = 3n^2 - 3n + 1.$$

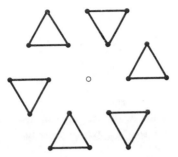

Fig. 3.24

The pattern starting $1 + 7 = 2^3$ is genuine. One way to see this is to see each hexagon as an array of cubes which will fit together to form a complete cube, as in Fig. 3.25. Another way is to spot that the formula for the nth hexagonal number can be written

$$3n^2 - 3n + 1 = n^3 - (n - 1)^3,$$

so that the first n hexagonal numbers can be written as

$$n^3 - (n - 1)^3$$
$$(n - 1)^3 - (n - 2)^3$$

$$\vdots$$

$$3^3 - 2^3$$
$$2^3 - 1^3$$
$$1^3 - 0^3.$$

When they are added up, the total is n^3.

Fig. 3.25

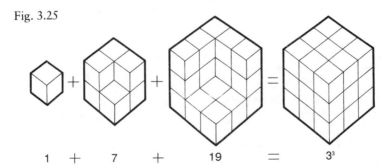

$$1 \quad + \quad 7 \quad + \quad 19 \quad = \quad 3^3$$

3E: *Faces and vertices*

Suppose that a regular polyhedron has F faces with n edges and vertices, with 3 faces per vertex. Then the polyhedron has $\frac{1}{2}nF$ edges and $\frac{1}{3}nF$ vertices. By Euler's relationship,

$$F + \tfrac{1}{3}nF = \tfrac{1}{2}nF + 2.$$

That is, $F = \frac{1}{6}nF + 2$, so that $6F - nF = 12$, or $(6 - n)F = 12$. Therefore F must be a factor of 12. $F = 2$ means $n = 0$, which makes no sense. $F = 3$ means $n = 2$, which also makes no sense. $F = 4$ implies $n = 3$, so there are four triangular faces, which is the regular tetrahedron. $F = 6$ implies $n = 4$, so there are six square faces, and this is the cube. Finally $F = 12$ implies that $n = 5$, and this is the regular dodecahedron with 12 pentagonal faces.

For 4 faces per vertex,

$$F + \tfrac{1}{4}nF = \tfrac{1}{2}nF + 2,$$

which implies that $n = 3$ and $F = 8$. This is the regular octahedron. For 5 faces per vertex,

$$F + \tfrac{1}{5}nF = \tfrac{1}{2}nF + 2,$$

which implies that $n = 3$ and $F = 20$. This is the regular icosahedron. We cannot have more than five faces per vertex because, calculating as above, we would conclude that there were fewer than three edges, which makes no sense.

3F: *Rigid frameworks*

Apart from the first two joints at the end of the initial bar, each joint added (with a pair of bars) creates one triangle. Therefore the total number of triangles is $n - 2$. The *total* number of bars meeting at all the vertices, counting each vertex separately, will be double the number of bars, since each bar joins two vertices. It is therefore $4n - 6$, and the average number of bars at each vertex is $4 - 6/n$.

This result might seem paradoxical, since a large tessellation of equilateral triangles, for example, clearly contains nearly 6 bars per vertex, only the vertices on the boundary having fewer. A partial explanation is that the tessellation of triangles is far from being just rigid. Figure 3.26 shows two ways in which the non-boundary portion of a large tessellation can have exactly 4 bars at each vertex. Suitably placed triangles on the boundary will ensure that the entire framework is rigid, and will also reduce the average number of bars per vertex to slightly under 4.

Note however that these 'solutions' cannot in fact be constructed by adding triangles in sequence, two bars and one new join at a time.

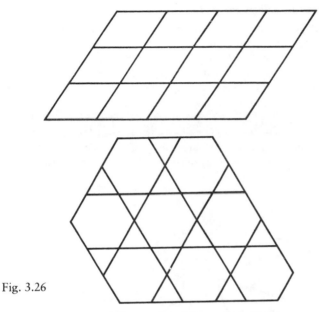

Fig. 3.26

An example of a large framework which is built up in that basic way, and is just rigid at every step, will be found on page 147.

3G: *Buckminsterfullerene and a familiar shape*

Figure 3.27 shows the shape, which can be obtained by *truncating* either a regular dodecahedron or a regular icosahedron. There are five hexagons round each pentagon, and pentagons and hexagons alternate round each hexagon. On the right is a modern football whose design copies the same shape exactly. (Old-fashioned footballs, in contrast, were made of almost-rectangular panels, six sets of three panels each, matching the faces of cube.)

Fig. 3.27

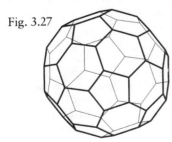 

3H: *Mountains, valleys and passes*

Figure 3.28 shows the original illustration next to a simplified sketch in which each mountain peak is a point, or vertex, each valley is a slight depression, and each ridge is marked by the line joining a pair of adjacent peaks, and separating two valleys. The entire surface of the asteroid will thus form a rough polyhedron to which Euler's relationship can be applied. So,

$$10 \text{ peaks} + 12 \text{ valleys} = \text{No. of ridges} + 2$$

and there are 20 ridges. This argument is quite general, and with careful modifications can be applied to maps of parts of the earth's surface.

Fig. 3.28

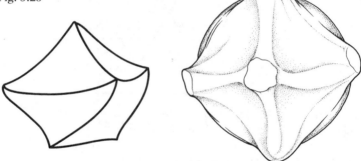

3I: *A polyhedral cheese*

The value of $V + F - E$ is called the *Euler characteristic* of a surface. It does not depend on how the surface is divided up into polygons, but only on the number and type of holes. For a sphere or any 'ordinary' polyhedron it is 2. For a doughnut or torus with one through-hole it is zero. For a solid with two through-holes it is -2, and for a solid with N such holes it is $-2(N - 1)$. For our solid, with 3 holes, it is -4, which fits the polyhedron on the left of Fig. 3.18, for which $V = 32$, $F = 28$ and $E = 64$.

3J: *Is this a number?*

Calculation will suggest the correct answer that $D^2 = D$. This suggests that D cannot be treated as a 'proper' number, because in ordinary algebra $D^2 = D$ implies that D is either 0 or 1. Also, $1 - D$ is . . . 09376, and E has the same property: $E^2 = E$. Finally, $D \times E = 0$, which again makes no sense if D and E are treated as ordinary numbers. The existence of D and E suggests the existence of *sequences* of digits which strongly resemble ordinary numbers but have quite different properties.

3K: *The reverse Koch snowflake*

The area removed from the inside of the original triangle, along one side, is identical in shape and size to the area added to the original triangle in the original construction (see Fig. 3.29). Therefore the reverse snowflake has lost the area that the normal snowflake gained. Since the area of the normal snowflake is $\frac{8}{5}$ of the original, the area of the reverse snowflake is $\frac{2}{5}$ of the original triangle.

Fig. 3.29

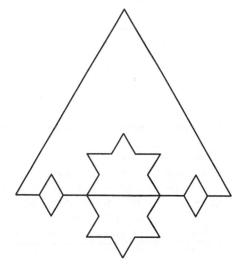

Chapter 4

The games of mathematics

A mathematician, like a painter or poet, is a maker of patterns. If his patterns are more permanent than theirs, it is because they are made with ideas ... a chess problem is *simply* an exercise in pure mathematics ... everyone who calls a problem 'beautiful', is applauding mathematical beauty ...

G. H. Hardy

Mathematics and abstract games like chess, draughts and the oriental game go have many features in common. Everyone agrees on the rules of chess and on the rules of arithmetic though they might have great difficulty in writing all the latter down (but then they might miss out one or two rules of chess also, yet still play the game correctly). Mathematicians agree on the rules of elementary Euclidean geometry, and on the rules of the many other well-known well-established kinds of mathematical system, though we must not forget the difficulties that mathematicians have had in the past with ideas such as 'polyhedron' and the trouble they still have with certain concepts that they use warily. Mathematics only strongly resembles abstract games when it has been developed scientifically to the stage where mathematicians are quite clear about what their ideas mean.

If you have a good memory, and an ability to visualize, you can play chess in your head, without an actual board, and you can also do all kinds of mathematics under the same conditions, as Hardy implies when he says that mathematics is made from *ideas*. Chess players recognize and applaud good play, from a single smart move to a brilliant combination to an entire game which is considered a masterpiece. Mathematicians recognize and applaud good mathematics, from clever tricks to brilliant proofs, and from beautiful conceptions to grand and deep ideas which advance our understanding of mathematics as a whole.

Positions at chess can be analysed and conclusions drawn with a very high degree of certainty, provided that the position is not too complicated. The same is true, with the same proviso, for mathematics – with the added condition that you must be quite certain that the 'rules of the game' are sound and non-contradictory. (Everyone takes for granted that the rules of chess are consistent: however, the rules of go have been adjusted more than once in this century by the Japanese professional go players' association to take into account anomalies such as the 'triple ko' which could result in a sequence of positions being repeated endlessly.)

It takes imagination and insight to discover the best moves, at chess or mathematics, and the more difficult the position, the harder they are to find. Chess players learn by experience to recognize types of positions and situations and to know what kind of moves are likely to be successful; they exploit brilliant local tactics as well as deep stategical ideas. So do mathematicians. Neither games nor mathematics play themselves – they both need a player with understanding, good ideas, judgement and discrimination to play them. To develop these essential attributes, the player must explore the game by playing it, thinking about it and analysing it. For the chess player and the mathematician, this process is scientific: you test ideas, experiment with new possibilities, develop the ones that work and discard the ones that fail. This is how chess players and mathematicians develop their tactical and strategical understanding; it is how they give *meaning* to chess and mathematics.

The mathematician and philosopher Bertrand Russell once said that 'mathematics may be defined as the subject in which we never know what we are talking about nor whether what we are saying is true.' Russell was mistaken. We become familiar with what we are talking about, and persuade ourselves of the truths of propositions, much as we explore the everyday world: if Russell was right, then we would be as ignorant of the rest of the world, not just of mathematics. (Talking of philosophy, mathematics quite naturally seems so real to most mathematicians, and is so widely shared throughout the world, that many mathematicians have persuaded themselves that it must exist in some special world of its own, much as the Greek philosopher Plato described, rather than just existing in mathematicians' heads, as chess exists in chess players' heads.)

Last but by no means least: millions of chess players enjoy playing chess who will never have the chance to play like Kasparov or Fischer or Alekhine, and millions of mathematicians enjoy mathematics who will never be a Hilbert or a Dirichlet or an Euler. There is also an important difference, to which we shall return later: most chess players spend zero time considering new rules or asking questions about the game, or inventing new games. They are satisfied with the game of chess as it is. Mathematicians are always asking questions *about* the mathematical games they are playing, and always inventing new games. They are seldom satisfied with any particular mathematical game as it is.

Traditional arithmetic is very much a simple game. We have a very clear idea of what numbers are, including fractions and irrational numbers, partly thanks to the Greeks, and the rules by which we operate with them are also clear. All we have to do is to decide what we want to do, and how to do it.

If it is only being used to cost the shopping at the supermarket, or work out an insurance claim, it may of course seem neither exciting nor imaginative. It is rather more interesting if we – for example – transform $8^3 - 7^3$ by this sequence of moves:

$$
\begin{aligned}
8^3 - 7^3 &= (7 + 1)^3 - 7^3 \\
&= 1 + 3 \times 7^2 + 3 \times 7 \\
&= 1 + 6 \times \tfrac{1}{2}(7^2 + 7) \\
&= 1 + 6 \times \tfrac{1}{2} \times 7 \times (7 + 1) = 1 + 6T_7,
\end{aligned}
$$

where T_7 is the 7th triangular number. This is the conclusion that we reached by sharp eyesight, in Chapter 2, and again in Chapter 3 by picturing the hexagonal numbers. By the above sequence of moves we can infer it by the rules of arithmetic. This inference suggests several important points, however.

The first is that the pattern of this sequence seems convincing; yet, can we be certain that exactly the same pattern of moves would work for, say, $9^3 - 8^3$ or $10^3 - 9^3$? If not, then we have as much reason to be suspicious of this sequence as we ever had to be suspicious of pattern spotting.

There is an answer to this difficulty: we switch to elementary algebra like this:

$$
\begin{aligned}
(x + 1)^3 - x^3 &= 1 + 3x^2 + 3x \\
&= 1 + 6 \times \tfrac{1}{2}(x^2 + x) \\
&= 1 + 6 \times \tfrac{1}{2}x(x + 1) \\
&= 1 + 6T_x.
\end{aligned}
$$

Each step works for this general x, and so our conclusion is, after all, extremely solid.

A second point is that, although this sequence of moves did not involve spotting any pattern, we still had to 'look ahead', as chess players say: we had to see where we were going, and make the appropriate moves to get there. You cannot do without *perception*, just because you are playing a game. Far from it, having a clear sight of the board, as it were, *seeing* the possibilities that are available, *looking ahead*, preferably several moves ahead ... are all essential. On the other hand, it has to be admitted that the amount of foresight required by this sequence of moves is absolutely minimal. It is what might be described by students as technique, or by experts as 'mere technique'.

'Technique' is a term used equally by chess players and mathematicians to describe sequences of moves which are standard, familiar and unoriginal. Once upon a time, the particular technique was an invention, a new discovery, but no longer. The *precise* sequence of moves required may never have been played before in the history of the world, yet no new ideas, no originality and no imagination are demanded, at least of the experienced player. (To the learner, of course, the most mundane sequences will appear novel and require original thought.)

Techniques can be elementary or advanced, but to the player who has mastered them, quite probably from studying a standard textbook, they come easily, which is not to say that they are unimportant or can be ignored. Quite the opposite – the most important techniques are the final results of some of the most significant breakthroughs in the history of mathematics, beautifully simplified and explained for the benefit of players everywhere. As such, technique is used by pure mathematicians and applied mathematicians alike, but in rather different ways. Pure mathematicians use familiar techniques on the way to discovering or proving new results; applied mathematicians use techniques to model phenomena in the real world, as we shall see in Chapters 9 and 10.

Here is an example of technique, with a little creativity, that is of interest to pure and applied mathematicians. Suppose that we have two sets of figures, and we have already calculated the average of each set separately. Under what circumstances will the average of all the figures together be equal to the average of the two averages we

---

Problem 4A *Which is the greater?*

If x and y are any two different positive numbers, which is larger, $x^2 + y^2$ or $2xy$?

already know? Suppose that the numbers in the first set are a, b, c, d ... and there are N of them, and that their average is P. Then

$$(a + b + c + d + \cdots)/N = P,$$

that is,

$$a + b + c + d + \cdots = NP.$$

Suppose similarly that the numbers in the second set are z, y, x, w, ... and that there are M of them with average Q, so that

$$(z + y + x + w + \cdots)/M = Q,$$

or

$$z + y + x + w + \cdots = MQ.$$

Then the total of all the numbers will be

$$a + b + c + d + \cdots + z + y + x + w + \cdots = NP + MQ,$$

and the total number of numbers will be $N + M$, so the average of all the numbers will be

$$(NP + MQ)/(N + M),$$

The average of the two original averages is just $\frac{1}{2}(P + Q)$, so we want to know when

$$\tfrac{1}{2}(P + Q) = (NP + MQ)/(N + M).$$

So far we have been arguing about what 'has to be' rather than using 'mere technique' but now the latter is just what we need. We simplify the last equation in the best schoolbook manner, without needing to think about the meaning of what we are doing, but just going through the motions, confident that we shall get the correct answer if we make no careless errors.

Problem 4B *Sums of square roots*

You are told that a, b and c are integers and that

$$\sqrt{a} + \sqrt{b} = \sqrt{c}.$$

What else can be said about a, b and c?

Problem 4C *Nicole Oresme and the harmonic series*

Nicole Oresme, Bishop of Lisieux (*c.* 1323–82) was the greatest European mathematician of the fourteenth century. He took a step in the direction of Copernicus by arguing that astronomical theory would be simplified if the earth were considered as having a daily rotation. He also shared with Copernicus the distinction of being one of the first economists; they both discussed the effects of the debasement of coinage on inflation.

Fig. 4.1

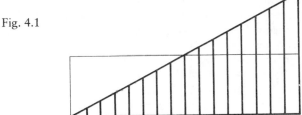

He used primitive coordinates to describe the position of a point, referring on a geographical analogy to the latitude and longitude of the point, and used this idea to demonstrate geometrically that if the speed of an object is increasing steadily, then its average speed is the average of its initial and final speeds. His diagram is shown in Fig. 4.1, where each vertical strip represents the speed at that time, increasing from left to right. The average speed is the horizontal bar in the middle.

He also demonstrated that the *harmonic series diverges*, that is, when added up for ever and ever, it does not tend to a limit. The series is

$$\tfrac{1}{2} + \tfrac{1}{3} + \tfrac{1}{4} + \tfrac{1}{5} + \tfrac{1}{6} + \tfrac{1}{7} + \tfrac{1}{8} + \cdots .$$

If we start adding up the series term by term, the sum increases very slowly. Yet it is true that by adding sufficient terms we can make the sum as large as we choose. How can we demonstrate this?

This is one of the great advantages of algebra, that many operations (not all, by any means) can be done more or less automatically, saving thought and effort:

$$(N + M)(P + Q) = 2(NP + MQ),$$
$$NP + NQ + MP + MQ = 2NP + 2MQ,$$
$$NQ - NP + MP - MQ = 0.$$

This might seem to be as 'simple' as we can get, but the N, M, P and Q are still all mixed up, and anyway by looking ahead just one step we can see that the last equation can be written

$$(N - M)(Q - P) = 0,$$

which is much more useful. Now the algebra is over, and we have to *interpret* our result. What does it mean? In this case the meaning is relatively simple. The product of $N - M$ and $P - Q$ is zero; so *either* $N - M = 0$, and so $N = M$, *or* alternatively $Q - P = 0$ and $Q = P$.

These are genuine alternatives. If either of them is the case, then the overall average will be the average of the original two averages. So there are two very different possibilities, either $N = M$ and the original two sets each have the same number of members; or $P = Q$, and the original two sets each have the same average.

*

Technique can be thought of as basic tactics. At the opposite extreme lie the dazzling tactics of a Kasparov or Tal, or an Euler or Ramanujan. Their tactics are highly creative, surprising, exquisite and profound, and usually hard to understand, though mathematics and chess share the feature that, provided you can play the game to a modest level, you can probably understand very brilliant tactics provided they are explained with a running commentary: this is the basis of chess columns in the newspapers and popular chess magazines.

That is what we shall do here: give a running commentary on some very simple results by two great masters. Here is a very simple example, starting with the series discovered by Euler:

$$\frac{1}{1^2} + \frac{1}{2^2} + \frac{1}{3^2} + \frac{1}{4^2} + \frac{1}{5^2} + \frac{1}{6^2} + \ldots = \frac{\pi^2}{6}.$$

Problem 4D   *Where does the symmetry come from?*

If we know that $a = b + c$, then we can deduce in a few well-chosen moves that

$$a^4 + b^4 + c^4 = 2b^2c^2 + 2c^2a^2 + 2a^2b^2.$$

Yet this latter expression is symmetrical in a, b and c, while the original equation was unsymmetrical. Where has the symmetry come from?

This series contains the squared reciprocals of all the natural numbers. We can find the same sum for the odd numbers only,

$$\frac{1}{1^2} + \frac{1}{3^2} + \frac{1}{5^2} + \frac{1}{7^2} + \cdots,$$

if we notice that, by multiplying the second series by the factor in the first bracket below, we get the original series:

$$\left(\frac{1}{1^2} + \frac{1}{2^2} + \frac{1}{4^2} + \frac{1}{8^2} + \frac{1}{16^2} + \cdots \right)\left(\frac{1}{1^2} + \frac{1}{3^2} + \frac{1}{5^2} + \frac{1}{7^2} + \cdots \right)$$

$$= \frac{1}{1^2} + \frac{1}{2^2} + \frac{1}{3^2} + \frac{1}{4^2} + \frac{1}{5^2} + \frac{1}{6^2} + \cdots.$$

It is a standard result, and also the result of a standard technique, that

$$1 + \frac{1}{2^2} + \frac{1}{4^2} + \frac{1}{8^2} + \frac{1}{16^2} + \cdots = 1 \Big/ \left(1 - \frac{1}{2^2}\right) = \frac{4}{3}.$$

Therefore,

$$1 + \frac{1}{3^2} + \frac{1}{5^2} + \frac{1}{7^2} + \cdots = \frac{\pi^2}{8}.$$

This is another result of Euler. The same idea can be used to sum, for example, the series in which instead of omitting all the even numbers, we omit all the multiples of 3:

$$1 + \frac{1}{2^2} + \frac{1}{4^2} + \frac{1}{5^2} + \frac{1}{7^2} + \frac{1}{8^2} + \frac{1}{10^2} + \frac{1}{11^2} + \frac{1}{13^2} + \cdots.$$

This series multiplied by

$$1 + \frac{1}{3^2} + \frac{1}{9^2} + \frac{1}{27^2} + \cdots = 1 \Big/ \left(1 - \frac{1}{3^2}\right) = \frac{9}{8}$$

gives the original series, so its sum is $\frac{8}{9} \times \frac{1}{6}\pi^2 = \frac{4}{27}\pi^2$. By using the same technique twice, we can sum of the series in which all multiples of 2 and 3 are omitted:

$$1 + \frac{1}{5^2} + \frac{1}{7^2} + \frac{1}{11^2} + \frac{1}{13^2} + \cdots.$$

*

Now let us move on to an example on the same theme from the works of the Indian genius Ramanujan.

Srinivasa Ramanujan was a poor Indian clerk when he wrote in 1913 to the distinguished English mathematician G. H. Hardy, sending in a letter the statements of more than one hundred theorems, without proofs. He had previously written to Baker and Hobson without getting any response. Hardy recognized his genius and arranged for him to visit Cambridge, where he workd until 1919 when, suffering from tuberculosis, he returned to India where he died at the age of 33.

Ramanujan's mathematics is like Tal's chess, dazzling. The following very simple example is adapted from the last paper he wrote in India before reaching England. (Despite the adaptation, however, it is as difficult as anything else in this book.) We start with the same series of Euler, and notice that it can be written as the product of an infinite sequence of series, like this:

$$1 + \frac{1}{2^2} + \frac{1}{3^2} + \frac{1}{4^2} + \cdots = \left(1 + \frac{1}{2^2} + \frac{1}{4^2} + \frac{1}{8^2} + \cdots\right)$$

$$\times \left(1 + \frac{1}{3^2} + \frac{1}{9^2} + \frac{1}{27^2} + \cdots\right)\left(1 + \frac{1}{5^2} + \frac{1}{25^2} + \cdots\right)$$

$$\times \left(1 + \frac{1}{7^2} + \frac{1}{49^2} + \cdots\right)\cdots\left(1 + \frac{1}{p^2} + \frac{1}{p^4} + \cdots\right)\cdots,$$

where, in the general factor, p represents an arbitrary prime number.

On the left is the sum of the squared reciprocals of the positive integers. On the right is a product of series, one for each prime, each being the sum of the even reciprocal powers of that prime. Each of these series can be summed, by a standard result, as we have seen:

$$1 + \frac{1}{2^2} + \frac{1}{4^2} + \cdots = \frac{1}{1 - 1/2^2}, \qquad 1 + \frac{1}{3^2} + \frac{1}{9^2} + \cdots = \frac{1}{1 - 1/3^2},$$

$$1 + \frac{1}{5^2} + \frac{1}{25^2} + \cdots = \frac{1}{1 - 1/5^2}, \cdots, 1 + \frac{1}{p^2} + \frac{1}{p^4} + \cdots = \frac{1}{1 - 1/p^2}, \cdots.$$

Therefore, we can write

$$\left(\frac{1}{1 - 1/2^2}\right)\left(\frac{1}{1 - 1/3^2}\right)\left(\frac{1}{1 - 1/5^2}\right)\cdots\left(\frac{1}{1 - 1/p^2}\right)\cdots$$

$$= 1 + \frac{1}{2^2} + \frac{1}{3^2} + \frac{1}{4^2} + \frac{1}{5^2} + \cdots = \frac{\pi^2}{6}. \tag{*}$$

Problem 4E   *A masterpiece by Euler*

One of Euler's most famous papers, famous not least because he eloquently explained how he achieved his results, involved this product:

$$(1 - x)(1 - x^2)(1 - x^3)(1 - x^4) \cdots.$$

The product goes on for ever, but the first few terms can be calculated by multiplying out. Thus the terms up to x^2 can be obtained from

$$(1 - x)(1 - x^2),$$

those up to x^3 from

$$(1 - x)(1 - x^2)(1 - x^3),$$

and so on. The resulting sequence starts:

$$1 - x - x^2 + x^5 + x^7 - x^{12} - x^{15} + \cdots.$$

Euler spotted a connection here with the *pentagonal numbers* which we have already met in Chapter 2. What is it?

Next we do the same, but starting with a different series sum, also due to Euler:

$$\frac{1}{1^4} + \frac{1}{2^4} + \frac{1}{3^4} + \frac{1}{4^4} + \frac{1}{5^4} + \ldots = \frac{\pi^4}{90}.$$

This we transform in the same way:

$$\left(\frac{1}{1 - 1/2^4}\right)\left(\frac{1}{1 - 1/3^4}\right)\left(\frac{1}{1 - 1/5^4}\right)\left(\frac{1}{1 - 1/7^4}\right) \cdots$$

$$= \frac{1}{1^4} + \frac{1}{2^4} + \frac{1}{3^4} + \frac{1}{4^4} + \ldots = \frac{\pi^4}{90}.$$

Ramanujan noticed here that each of the bracketed expressions, such as $1 - 1/5^4$ is the *difference of two squares*. In other words, that

$$\frac{1}{1 - 1/5^4} = \frac{1}{(1 - 1/5^2)(1 + 1/5^2)}.$$

Factorizing all the bracketed expressions, we get:

$$\frac{1}{(1 - 1/2^2)(1 + 1/2^2)} \cdot \frac{1}{(1 - 1/3^2)(1 + 1/3^2)} \cdot \frac{1}{(1 - 1/5^2)(1 + 1/5^2)} \cdots$$

$$= \pi^4/90.$$

We can now divide this equation by the equation (∗). The result is a similar product, but with all the minus signs replaced by pluses:

$$\left(\frac{1}{1 + 1/2^2}\right)\left(\frac{1}{1 + 1/3^2}\right)\left(\frac{1}{1 + 1/5^2}\right)\cdots = \frac{\pi^2}{15}.$$

Now we work backwards. Each of these expressions, with a plus instead of a minus, is also the sum of a series. In fact each series is very similar to the original series, the only difference being that the signs now alternate:

$$\left(1 - \frac{1}{2^2} + \frac{1}{4^2} - \frac{1}{8^2} + \cdots\right)\left(1 - \frac{1}{3^2} + \frac{1}{9^2} - \frac{1}{27^2} + \cdots\right)$$

$$\times \left(1 - \frac{1}{5^2} + \frac{1}{25^2} - \cdots\right)\cdots = \frac{\pi^2}{15}.$$

What happens when we multiply out these bracketed series? The terms will be identical to the original series except for the difference made by the changes of sign. What will these changes be?

If a term is formed by the product of an even number of terms within the bracketed expressions, it will still appear with a positive sign; but, if it is the product of an odd number of terms, its sign will be negative. So the product is

$$1 - \frac{1}{2^2} - \frac{1}{3^2} + \frac{1}{4^2} - \frac{1}{5^2} + \frac{1}{6^2} - \frac{1}{7^2} - \frac{1}{8^2} + \frac{1}{9^2} - \cdots = \frac{\pi^2}{15}.$$

Ramanujan finally subtracted this equation from the first equation and halved the result:

$$\frac{1}{2^2} + \frac{1}{3^2} + \frac{1}{5^2} + \frac{1}{7^2} + \frac{1}{8^2} + \frac{1}{11^2} + \frac{1}{12^2} + \frac{1}{13^2} + \frac{1}{17^2} + \frac{1}{18^2} + \cdots$$

$$= \frac{1}{2}\left(\frac{\pi^2}{6} - \frac{\pi^2}{15}\right) = \frac{\pi^2}{20}$$

In this series, not merely the multiples of 2, or the multiples of 3, have been deleted from the natural numbers in the denominators: all the natural numbers with an even number of prime factors have been deleted, a far harder task.

Ramanujan was a brilliant tactician, who also had a wonderful eye for observing patterns. He was a superb mathematical scientist, with a special ability to spot analogies and relationships. There is a natural connection between these two abilities, which Gauss and Euler also demonstrated. In a well-understood game, tactical calculation, of a mundane sort, is relatively easy, so it is very easy to generate large numbers of examples, which in turn provide meat and drink for the scientist looking for patterns. This is as true of elementary geometry as it is of algebra or the properties of numbers. It has often been said that most of elementary Euclidean geometry was first discovered by scientific experiment, and only then proved by reasoning.

If you are able to calculate, but you fail to spot a pattern, then of course you may be in trouble, as the next example amusingly shows. In 1869, Landry (the same Landry who factorized $2^{64} + 1$) announced that

$$2^{58} + 1 = 5 \times 107{,}367{,}629 \times 536{,}903{,}681$$

and he explained:

No one of the numerous factorizations of the numbers $2^n \pm 1$ gave as much trouble and labour as that of $2^{58} + 1$. This number is divisible by 5; if we remove this factor, we obtain a number of 17 digits whose factors have 9 digits each. If we lose this result, we shall miss the patience and courage to repeat all calculations that we have made and it is possible that many years will pass before someone else will discover the factorization of $2^{58} + 1$.

It is typical of the mystery of mathematics, that while it is very easy to multiply two numbers together, by hand using pencil and paper if necessary, the reverse process of finding the factors of a number is usually very difficult.

Persistence and courage were indeed needed in those days, before modern calculating machines had been invented. However, in this case, Landry had failed to notice a smart move which would have saved him most of his trouble. Only a few years later, Aurifeuille noticed that Landry's third factor $536{,}903{,}681$ was almost exactly 5 times the second factor $107{,}367{,}629$. In fact, the difference is

$$536{,}903{,}681 - 5 \times 107{,}367{,}629 = 65{,}536 = 2^{16}.$$

By introducing half this power of two, $2^{15} = \frac{1}{2} \times 2^{16}$, Aurifeuille worked out that

$$2^{58} + 1 = (2^{29} - 2^{15} + 1)(2^{29} + 2^{15} + 1),$$

which can be seen by expanding the right-hand expression, when all but two terms cancel out. Had Landry realized this, all his calculations could have been saved. However, it has to be admitted that Aurifeuille did use his scientific observation of Landry's result to find this pattern.

Later, Eduard Lucas, who invented the Tower of Hanoi and other mathematical recreations, pointed out that $2^{58} + 1$ happens to be the sum of two squares in two different ways:

$$2^{58} + 1 = (2^{29})^2 + 1^2 = (2^{29} - 1)^2 + (2^{15})^2.$$

Taking the average of the two forms, Lucas noticed that

$$2^{58} + 1 = \tfrac{1}{2}[(2^{29})^2 + 1^2 + (2^{29} - 1)^2 + (2^{15})^2]$$
$$= (2^{29} - 2^{15} + 1)(2^{29} + 2^{15} + 1),$$

and that, in the general case,

$$2^{4n+2} + 1 = (2^{2n+1} - 2^{n+1} + 1)(2^{2n+1} + 2^{n+1} + 1).$$

Lucas also noticed that you can get this pattern by the simple move of replacing x by 2^n in the following pattern:

$$4x^4 + 1 = (2x^2 - 2x + 1)(2x^2 + 2x + 1).$$

Aurifeuille's formula is the special case when $n = 14$, or $x = 2^{14}$.

Lucas's argument, unlike Aurifeuille's, did not depend on scientific observation at all. Lucas was just playing the game of elementary algebra, rather well. Lucas spotted some very good moves, but not by collecting data like a naturalist and looking for patterns. He was looking ahead and using his imagination, like a good chess player.

As Professor Hobson said, at a meeting of the British Association for the Advancement of Science, in 1910:

The possession of a body of [objects and rules] would be ... quite insufficient for the development of a mathematical theory. With these alone the mathematician would be unable to move a step. In the face of an unlimited number of possible combinations, a principle of selection of such as are of interest, a purposive element, and a perceptive faculty are essential for the development of anything new.

Hobson went on to compare a mathematical argument to a musical melody, which is much more than the sum of the individual notes that compose it. This is an appropriate metaphor, not least because some mathematicians have produced mathematical 'melodies' as Mozart produced his music, or Capablanca played chess – apparently effortlessly, as if the ideas flowed out of their heads with no intervention on their part. (Though we know from examination of Mozart's

Problem 4F   *Find the factors*

It is difficult to find the factors of large numbers, but it also may not be easy to factorize algebraic expressions. Consider, for example,

$$x - 1 = x - 1$$
$$x^2 - 1 = (x - 1)(x + 1)$$
$$x^3 - 1 = (x - 1)(x^2 + x + 1)$$
$$x^4 - 1 = (x - 1)(x + 1)(x^2 + 1)$$
$$x^5 - 1 = (x - 1)(x^4 + x^3 + x^2 + x + 1)$$
$$\vdots$$

This pattern suggested to mathematicians that – perhaps! – in any irreducible factor of $x^n - 1$ that has only integral coefficients, each of those coefficients must be 0 or ± 1. (Here, an *irreducible* factor is one that will not factorize further into factors with integral coefficients.) Finally, after many years of fruitless attempts to prove that this was so, it was discovered that it was in fact false. One of the factors of

$$x^{105} - 1$$

actually has coefficients that break this rule.

Here are two simpler problems: what are the factors of these two expressions?

$$1 + x^4 + x^5, \qquad 1 + x^7 + x^8.$$

autographs, or Capablanca's or Euler's own comments, that more effort is required than we see in the finished work of art.)

Where does this imagination and insight, this fluency and confidence, come from? How is Hobson's 'perceptive faculty' developed? It is a mixture of natural talent – whence the Fischers and Eulers of chess and mathematics – and experience. The player learns by experience, by exploration, by the *scientific* exploration of the game, just as if the game were a miniature world to be explored and understood, like the 'real' world in which we live our everyday lives.

By practice and experience, much of it vicarious, through studying the achievements of others, we develop the strength to tackle novel and unfamiliar situations.

*

Problem 4G *Summing a series*

The following infinite series rings a bell; the coefficients are just the triangular numbers:

$$1 + 3x + 6x^2 + 10x^3 + 15x^4 + \cdots$$

How will multiplying the series by $1 - x$ help to find its sum?

Fortunately for the vast majority of mathematicians, tactical brilliance at the level of Ramanujan is not necessary to do mathematics. Most mathematics involves a combination of more-or-less original tactics plus strategical ideas, as the next case illustrates. While looking, in Chapter 2, at the sequence of squares,

$$1 \quad 4 \quad 9 \quad 16 \quad 25 \quad 36 \quad 49 \quad 64 \quad 81 \quad 100 \quad \cdots,$$

we noticed that the differences are the odd numbers in sequence, 1, 3, 5, 7 . . . . Another way of saying this is to say that every odd number is the difference between two squares.

So are many even numbers: for example, $8 = 9 - 1$ and $40 = 49 - 9$. Some even numbers, however, are not the difference between two squares. The number 6 is not, for example, and neither is 10, or 18, or 22. In fact, a scientific test suggests that any number which is even, but not a multiple of 4, cannot be the difference of two squares. Really? Why not? At this point we will stop behaving like scientists and tackle the problem by playing the game of elementary algebra.

Suppose that a number N is the difference between two integer squares:

$$N = p^2 - q^2.$$

How might we find the two numbers, p and q? Experience has taught the mathematician who sees the 'difference of two squares', especially in a situation like this, that a good move is almost certainly to *factorize* them. This is a typical example of the lessons of experience. Whatever the values of p and q,

$$p^2 - q^2 = (p + q)(p - q).$$

So we know that

$$N = (p + q)(p - q).$$

This statement now requires a little *interpretation*. We need to look at it, and 'see what it says': that N is the product of two factors, one of which is the sum of two numbers, p and q, and the other their difference. If we call these factors of N, X and Y, in that order, then

$$N = XY \quad \text{and} \quad X = p + q \quad \text{and} \quad Y = p - q$$

This is now looking promising. Why? Because if we know the values of X and Y, then we can solve these equations and find the values of p and q, and we will then know how to write N as the difference of two squares.

So, adding the equations: $\quad X + Y = 2p \quad \text{and} \quad p = \frac{1}{2}(X + Y)$.

Subtracting: $\quad X - Y = 2q \quad \text{and} \quad q = \frac{1}{2}(X - Y)$.

So, the problem is almost solved. We take our number N, we split it into two factors, X and Y, and then we calculate p and q from these formulae. What could go wrong? There is only one possible fly in the ointment. It might just be that, however we split N into two factors, one of them is always even and the other is always odd. If *that* happened, then $X + Y$ and $X - Y$ would both be odd; then, when we halved them, p and q would not be whole numbers.

When might this difficulty occur in practice? If N is odd, it will certainly not happen, because all the factors of N will be odd. If N is a multiple of 4, then it need not happen, because we can always arrange for each of the two factors to be a multiple of 2, and that's fine. The difficulty is only insuperable when N is a multiple of 2, but not of 4. For example, when $N = 10$. Whether we write $10 = 10 \times 1$ or $10 = 5 \times 2$, one factor is odd and the other even, and p and q will not be integers.

Well, well, well! So our suspicions were correct – but our new found confidence does not come from testing, testing, testing, and hoping that any suspicions we might have are unjustified, but from an argument which applies to any number N, which does not need to be repeated again and again, and which also explains very clearly *why* numbers like 18 are not the differences of two squares.

As an added bonus, we have almost solved the extra problem of saying *in how many ways* a particular number is the difference of two squares. This is a typical outcome. We are now quite certain of what we only strongly suspected before, and we are no longer in the dark, because we understand *why* the pattern exists – it is no longer a mystery. Yet this result was not achieved through tactical adroitness alone. We had to look very closely at the algebraic statement $N^2 = (p + q)(p - q)$ and interpret it, decide what it meant, 'see it in a new way'. Only then could we make further progress.

There is another difference which you may also have spotted. Playing the game of mathematics is much harder than investigating scientifically! To jot down some numbers, a few differences, and spot a pattern is child's play compared to playing the game of algebra.

Although this is a rather simple game, as the game of algebra goes, we still needed at least two smart ideas, and there was nothing to put these ideas into our heads except past experience and imagination.

As Gauss wrote, 'A great part of [number theory] derives an additional charm from the peculiarity that important propositions, with the impress of simplicity upon them, are so often easily discoverable by induction, and yet are of so profound a character that we cannot find their demonstration till after many vain attempts . . .'. If the great Gauss had such feelings, no wonder if we do! Our next example illustrates this difficulty very well. Readers who wish to are invited to skip at once to page 101.

The problem is to prove that it is not possible for the sum *and the difference* of two perfect squares both to be squares. In other words,

$$x^2 + y^2 = r^2 \quad and \quad x^2 - y^2 = s^2$$

is not possible in integers. We are following Fermat's own argument, slightly adapted. We start by assuming that such an x and y exist and have no common factor. This is fair, because if they did, then the same factor would have to divide r and s, and we could divide all four numbers by the common factor, and start all over again, until we have reduced the problem to one where no such factorization is possible.

Problem 4H *The three Dutchmen and their wives*

'There came three Dutchmen of my acquaintance to see me, being lately married; they brought their wives with them. The men's names were Hendrick, Claas, and Cornelius; the women's Geertrick, Catriin, and Anna; but I forget the name of each man's wife.'

'They told me that they had been at market, to buy hogs; each person bought as many hogs as they gave shillings for each hog; Hendrick bought 23 hogs more than Catriin, and Claas bought eleven more than Geertrick; likewise, each man laid out 3 guineas more than his wife. I desire to know the name of each man's wife?' (A guinea was 21 shillings.)

Problem 4I   *Logic and mathematics*

This conundrum first appeared in an old Civil Service examination. It proved so popular that the *New Statesman* persuaded the problemist Hubert Phillips to set similar problems to take the place of crossword and bridge columns.

'The driver, fireman and guard of a certain train were Brown, Robinson and Jones, and the passengers included Mr Brown, Mr Robinson and Mr Jones. Mr Robinson lived in Leeds; the guard lived midway between Leeds and London. Mr Jones' income is £400 2s 1d, and the guard's income is exactly one-third of his nearest passenger neighbour. The guard's namesake lives in London. Brown beats the fireman at billiards. What is the name of the engine-driver?'

(£1 = 20s = 240d.)

Next, we settle the question of which of x, y, r and s are odd or even. We have already decided that x and y cannot both be even, with a common factor 2. Now we recall that the square of an odd number is always of the form $4n + 1$, while an even square is a multiple of 4. Therefore, if x and y are both odd, r^2 will be of the form $4n + 2$, and cannot be a square. So x and y are not both odd. Therefore one is odd and one is even.

If x is even and y is odd, then $x^2 - y^2$ will be of the form $4n - 1$ and cannot be a square. Therefore x is the odd number and y is even, and so r and s are both odd.

So far, so good. The argument has been quite cunning, but not very cunning. In fact, readers may have noticed that the argument has been very similar indeed to the kind of arguments needed to solve those popular logical puzzles about the driver, the fireman and guard – see Problem 4I!

Now for Fermat's truly brilliant move. He defines u and v by the equations

$$u = \tfrac{1}{2}(r - s) \text{ and } v = \tfrac{1}{2}(r + s)$$

(u and v are integers since r and s are both odd), and then points out that, by calculation,

$$u^2 + v^2 = x^2.$$

This seems to say that x^2 is the sum of two squares – which does not seem very clever, because we knew at the start that $x^2 - y^2 = s^2$, from which it follows that $x^2 = y^2 + s^2$, which appears to say the same thing!

Not so, because u and v have an extra property that y and s do not possess. This is the fact that, again by calculation,

$$\tfrac{1}{2}uv = \tfrac{1}{4}y^2.$$

Since y is even, $\tfrac{1}{4}y^2$ is a perfect square, and so $\tfrac{1}{2}uv$ is a perfect square.

Now Fermat goes back to the equation, $u^2 + v^2 = x^2$ and uses the fact that u, v and x are a Pythagorean triple and that u and v have no common factor. From these facts he deduced that x is the sum of two squares:

$$x = X^2 + Y^2,$$

and one of u or v is the difference of the same squares, $X^2 - Y^2$, and the other is the product, $2XY$. It doesn't matter which is which because Fermat is only interested in the product $\tfrac{1}{2}uv$ which he already knows is a perfect square. In other words,

$$\tfrac{1}{2}uv = \tfrac{1}{2}\cdot 2XY\cdot(X^2 - Y^2) = XY(X^2 - Y^2) = XY(X + Y)(X - Y)$$

is a perfect square. At this point we are near the finishing line. Fermat now observes that X and Y have no common factor because, if they had a common factor so would x, and u, and v, and therefore so would r and s, and we eliminated that possibility right at the start. But, in that case, $XY(X + Y)(X - Y)$ is the product of four factors with no common factors, and is also a perfect square. It follows that X and Y and $X + Y$ and $X - Y$ are all individually perfect squares.

How does this help? It helps because here is another example of Fermat's *method of infinite descent* which we met briefly in Chapter 2. Fermat started with the perfect squares of x^2 and y^2 whose sum and difference were also, hypothetically speaking, perfect squares. From this hypothesis, he has inferred that there is another pair of perfect squares, X and Y, with the same property *and X and Y will be smaller than x and y.*

This conclusion sets the method of infinite descent in motion. If X and Y really exist, then the same process will lead to another pair of perfect squares, which are smaller still, with the same property. And so on, and on, and on . . . endlessly. But this is not possible, because there are only a limited number of integers less than the original x^2 and y^2. This contradiction proves that the original assumption is false. Therefore the sum and difference of two squares cannot both be squares.

*

Wonderful! This argument is a great masterpiece in the history of mathematics. It is also the only proof of Fermat's which is extant. Although Fermat often claimed to have proved the theorems he stated, he never communicated the proofs, with this one exception, though he did imply that many of them depended on the 'method of infinite descent'.

It is that method, as much as his tactical brilliance, which makes this proof so notable. Fermat's novel method is the first recognition of a property of the positive integers that no one had previously

---

Problem 4J *Fermat's Last Theorem*

Fermat is best known for his 'last theorem', which has been neither proved nor disproved for 350 years. Fermat had a copy of Bachet's translation of the works of the Greek algebraist Diophantos, whose problems provided much of Fermat's inspiration. Problem 8 of Book II was 'To divide a given square number into two squares.' Fermat wrote in the margin these famous words, 'To divide a cube into two cubes, a fourth power, or in general any power whatever into two powers of the same denomination above the second is impossible, and I have assuredly found an admirable proof of this, but the margin is too narrow to contain it.'

Whether Fermat really had a proof of this theorem is doubted by almost all mathematicians. In the meantime, the efforts of some of the greatest mathematicians to solve it have led to remarkable advances in number theory and algebra, without answering the original question. The latest attempt occurred in the summer of 1993 when, in a blaze of publicity, Professor Andrew Wiles, British born but working at Princeton University, announced towards the end of a long lecture, almost as an afterthought, that he had proved it. His 'proof' also introduced some novel ideas, and was initially thought to be sound. Once it had been examined by the few mathematicians able to understand it, a flaw appeared, and Professor Wiles withdrew his claim. He is expected to 'fill the gaps' very soon.

However, it is known that Fermat's theorem is true for all small powers. In particular, it is true for fourth powers:

$$x^4 + y^4 = z^4 \quad \text{is impossible in integers.}$$

How can this conclusion be deduced from Fermat's own theorem that $x^2 + y^2$ and $x^2 - y^2$ cannot both be squares?

noticed, or considered important. The fact that there are only a finite number of positive integers less than any given positive integer seems obvious when it is pointed out, and it probably seems equally obvious that it is not a very useful fact. Fermat demonstrated otherwise. It is an essential feature, which the fractions, for example, do not share. Given any positive fraction, no matter how small, there are any number of smaller positive fractions. Like the Greeks who discovered, to their distress, that most square roots are irrational, Fermat had discovered an absolutely basic fact about numbers, which no one else had noticed – and then exploited it brilliantly. A rare achievement!

The games of mathematics do not include only arithmetic and various kinds of algebra. Far from it. Geometry is often much like a game, and the traditional geometry of Euclid is especially game-like.

The omission which we noted in Chapter 3 and which Hilbert repaired was never a serious block to the discovery or proof of theorems, which started with the basic principles laid down by Euclid at the start of his *Elements*, and built up, theorem upon theorem, to create a mathematical edifice which provoked the admiration and the envy of philosophers and moralists, who only wished that their confused and convoluted subjects could be as elegantly and simply and powerfully constructed from first principles.

Without going right back to Euclid's starting points, let's look at some examples of this elegant construction. Figure 4.2 shows P as a variable point joined to two fixed points A and B on the circle. Experiment with actual drawings and a protractor will easily reveal that the angle $\widehat{APB}$ is constant, provided that P stays on the longer arc AB, within the limits of reasonable experimental error.

We wish to prove that this is so, by using, as it were, the *rules of the game*. Our initial clue lies in the idea of a circle, which is the set

Fig. 4.2

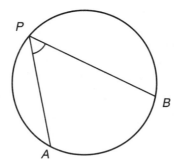

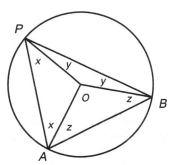

Fig. 4.3

of all points equidistant from a fixed point, its centre. So, marking the centre as O, we know by definition that OA, OB and OP are equal in length.

This suggests a theorem which is proved early in Euclid's *Elements*, that in an isosceles triangle (a triangle with two equal sides) the base angles are equal. Completing the isosceles triangles (Fig. 4.3), we can see that the angles marked x, y and z are equal in pairs. Does this help? Yes, because we know that the angle sum of a triangle is constant; so

$$2x + 2y + 2z \text{ is constant}$$

and therefore $x + y + z$ is constant. But z is constant anyway, because the centre O and the points A and B are not moving.

Problem 4K   *The angle in a semicircle*

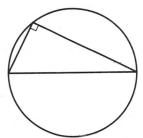

Fig. 4.4

The Babylonians knew that the 'angle in a semicircle' is a right angle. In other words, if the ends of a diameter of a circle are joined to any point on the circle, the two lines are perpendicular. How can this fact be deduced from the theorem that 'the angle in the same segment is constant'?

Therefore $x + y$ is constant and that is just the angle at P. This is known as the 'angle in the same segment' theorem. As it happens, and not surprisingly, the converse is also true: if A and B are fixed points, and a point P moves so that $\widehat{APB}$ is constant, then the path of P is an arc of a circle through A and B.

This simple problem and argument are typical of elementary geometry. The argument is highly visual, of course, but although the figure is there 'in front of our eyes', it is not necessarily immediately obvious what to do next, even in an extremely simple case like this. The fact is that in elementary geometry there are usually far more possible moves to consider than in elementary algebra, and the best move can be very obscure indeed.

This is illustrated by a story told by Roger Penrose, the mathematician and physicist and creator of twistor theory. He once found that in his work he needed to use Ptolemy's theorem. He knew what the theorem said, of course, but he decided (as any mathematician might) that, while he was about it, he would prove this theorem for himself. To his surprise, it was much harder to prove than he expected. So he looked up Ptolemy's original proof. What he found is illustrated in Fig. 4.5. The theorem says that, in any quadrilateral whose vertices lie on a circle, as in the left-hand figure,

$$AB \times CD + BC \times DA = AC \times BD.$$

Ptolemy proved this by drawing the line BE, in the right-hand figure, so that angle $\widehat{ABE}$ = angle $\widehat{DBC}$, and triangle BEA is similar to triangle BCD.

How did Ptolemy know that this bizarre extra line would lead to a proof? We do not know. We can follow Ptolemy's reasoning, once the idea has been spotted, but Ptolemy's original motivation is lost forever. It is often the case in geometry that there are so many

Fig. 4.5

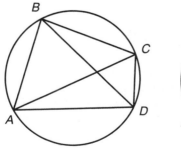
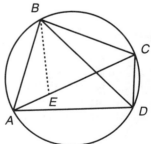

Problem 4L   *A special case of Ptolemy's theorem*

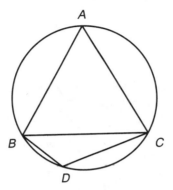

Fig. 4.6

Figure 4.6 shows an equilateral triangle *ABC*, and *D* is any point on the arc *BC*. What neat theorem does Ptolemy's theorem lead to in this special case?

possible lines of argument, so many possible constructions, that we are embarrassed to find a suitable one, and that success comes through unusual and unexpected means.

Having deduced one basic result about angles in segments of circles, we can now more easily deduce another result. In Fig. 4.7, a cyclic quadrilateral has its vertices joined to the centre *O*. By the same argument as before, here are four isosceles triangles, and the sum is

$$2p + 2q + 2r + 2s = 360°$$

Fig. 4.7

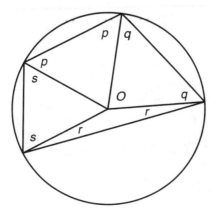

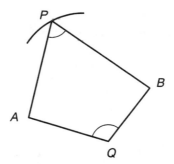

Fig. 4.8

because the angles of any quadrilateral sum to 360°. So $p + q + r + s = 180°$ and so the sum of either pair of opposite angles of the cyclic quadrilateral sums to 180°. Going back to our original figure, and adding a variable point Q on the shorter arc AB, we can now say that angle P is constant, and angle Q is constant, and their sum is 180°.

This result also has a converse (see Fig. 4.8). If A, B and Q are fixed, and P varies so that angle P + angle Q = 180°, then P lies on the circle through A, Q and B.

The figure of one circle and three points on it could hardly be simpler. The next figure (Fig. 4.9) is more complicated.

Three circles have a common point, C. The vertices P and Q lie one on each circle, and the sides of the triangle apparently pass through the remaining intersections of the circles, X, Y and Z. It appears that PY and QX meet at R, on that circle. Experiment will confirm that this is always so, provided only that the circles share a common point.

Fig. 4.9

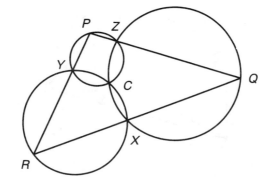

Problem 4M *A constant chord*

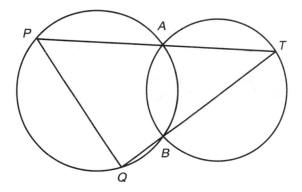

Fig. 4.10

Two fixed circles meet at A and B, and T moves along the major arc AB (see Fig. 4.10). Why will the length of the chord PQ be constant as T changes position?

It means that as P moves round the left-hand circle and Q moves round the right-hand circle, R will move round the lower circle. Another way to describe the situation is to say that there are an infinite number of triangles with their vertices on the circles and their sides passing through the remaining intersections of the same circles. Why should this be?

From the previous theorem about circles, angle P is constant and so is angle Q. But the angle sum $P + Q + R$ is also constant, so angle R must be constant. Therefore, as the points P and Q vary in position, the point R lies on an arc of a circle through X and Y.

This does not explain why R lies on a circle through C, so let's join X, Y and Z to C, in order to exploit our second result (Fig. 4.11). We have

$$P + Q + R = 180°, \qquad P + \widehat{YCZ} = 180°, \qquad Q + \widehat{ZCX} = 180°,$$

$$\widehat{YCZ} + \widehat{ZCX} + \widehat{XCY} = 360°,$$

from which it follows that $R + \widehat{XCY} = 180°$. In other words, the point R does lie on the circle through X, C and Y.

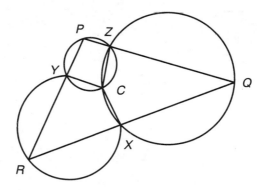

Fig. 4.11

This theorem can be generalized to more than three circles. Figure 4.12 shows four circles through a common point. Four lines through the outer intersections of the circles form the the sides of a quadrilateral whose vertices lie one on each circle.

The three-circle theorem can be seen in a slightly different way. Instead of starting with three circles, we start with the triangle, any triangle ABC, and mark one point on each side, A' opposite A, etc.

Fig. 4.12

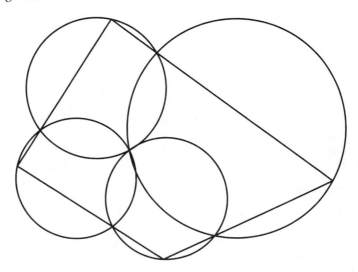

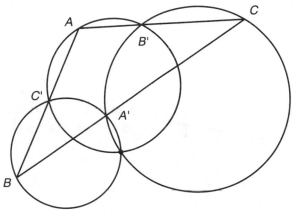

Fig. 4.13

We then draw the circles $AB'C'$, $BC'A'$ and $CA'B'$ and find that they have a common point (Fig. 4.13). In this form, it is called the pivot theorem, and it has an analogue in three dimensions; see Fig. 4.14. Take any tetrahedron and mark one point on each of the six edges. Then the four spheres through $AYVU$, $BWZU$, $CXVW$ and $DXZY$ all share a common point.

It is typical that there is more than one way of looking at a geometrical figure, just as there are many ways of looking at lines of algebra. Perception, 'seeing', is an essential feature of mathematics. This is obvious when we are looking for patterns – how can you possibly 'spot' a pattern if you cannot in some sense 'see' it? But it is just as true when the mathematician is looking for hidden connections,

Fig. 4.14

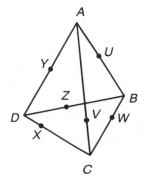

Problem 4N   *Touching circles*

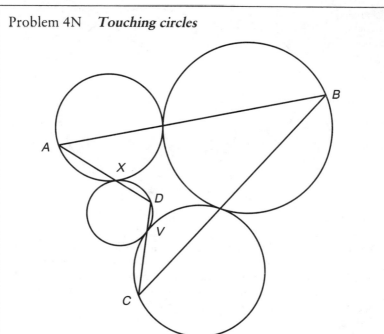

Fig. 4.15

Figure 4.15 shows four externally touching circles, and the quadrilateral formed by four points on the circumferences with sides passing through the tangent points. If vertex A is marked first, and then vertices B and C, why will the lines AX and CV meet on the fourth circle?

or studying a position in a mathematical game, searching for a tactical sequence, or trying to 'see' the possibilities clearly. Superficially, it might seem that it is only geometry (and related fields of mathematics) that depends on perception, but this is not so. Perception is everywhere in mathematics.

As it happens there is yet another way of looking at the pivot theorem, where now Fig. 4.13 should be thought of as a two-dimensional slice through a three-dimensional model. Imagine three spheres S_1, S_2, S_3 with just one point in common. Take points A on S_1 and B on S_2 such that the line AB passes through a point of the circle common to S_1 and S_2. Then there will be a plane slice of the spheres which includes the line AB and the common point of all three spheres. In this slice, we can complete the figure for the pivot theorem, and find a point C on S_3 such that AC and BC pass through points on the circles common to S_1 and S_3, and S_2 and S_3. In this case, point A does not just wander freely round a circle, but even more freely over the surface of a sphere, while B and C do likewise.

Solutions to problems, Chapter 4

4A: *Which is the greater?*

The square $(x - y)^2$ will certainly be positive, but it is equal to $x^2 + y^2 - 2xy$. Therefore, $x^2 + y^2$ is greater than $2xy$.

By replacing x^2 by X and y^2 by Y, this can be transformed into the statement that $X + Y > 2\sqrt{(XY)}$, or $\frac{1}{2}(X + Y) > \sqrt{(XY)}$, which says that the *arithmetic mean* of two unequal positive numbers, on the left, is greater than their *geometric mean*, on the right.

4B: *Sums of square roots*

Let $\sqrt{a} + \sqrt{b} = \sqrt{c}$. Then squaring gives $a + b + 2\sqrt{(ab)} = c$, so $2\sqrt{(ab)} = c - a - b$ is an integer. Then $\sqrt{(ab)}$ must be an integer, and the product ab must be a perfect square. This is also a sufficient condition, because we can work backwards, introducing c as $c = a + b + 2\sqrt{(ab)}$.

4C: *Nicole Oresme and the harmonic series*

Oresme bracketed the terms like this,

$$\frac{1}{2} + \left(\frac{1}{3} + \frac{1}{4}\right) + \left(\frac{1}{5} + \frac{1}{6} + \frac{1}{7} + \frac{1}{8}\right)$$

$$+ \left(\frac{1}{9} + \frac{1}{10} + \frac{1}{11} + \frac{1}{12} + \frac{1}{13} + \frac{1}{14} + \frac{1}{15} + \frac{1}{16}\right) + \cdots,$$

where the successive groups of terms contain 1, 2, 4, 8, 16, 32, . . . , 2^n, . . . terms, and noted that the terms in each additional bracket add up to more than $\frac{1}{2}$. So, by taking as many brackets as we need, we can make the total as large as we please. This takes many terms, however, because the number of terms in each bracket is doubling from one bracket to the next.

In fact, it takes the first 272,400,600 terms to pass 20; their sum is approximately 20.00000 00016.

4D: *Where does the symmetry come from?*

We deduce the result by this calculation: since $a = b + c$, we have

$$a - b - c = 0.$$

Squaring gives

$$a^2 + b^2 + c^2 + 2bc - 2ca - 2ab = 0.$$

Separating and squaring again yields

$$(a^2 + b^2 + c^2)^2 = (2ac + 2ab - 2bc)^2,$$

so

$$a^4 + b^4 + c^4 + 2b^2c^2 + 2c^2a^2 + 2a^2b^2$$
$$= 4a^2c^2 + 4a^2b^2 + 4b^2c^2 + 8a^2bc - 8ab^2c - 8abc^2.$$

But

$$8a^2bc - 8ab^2c - 8abc^2 = 8abc(a - b - c) = 0$$

and the result follows. However, as calculation confirms, we would have reached the same result if we had started with $b = a + c$, or $c = a + b$, or the equation $a + b + c = 0$. They all lead to the same final equation because the effects of the initial differences in sign are eliminated when we square and square again. Therefore we can conclude that $a - b - c$, $b - c - a$, $c - a - b$, and $a + b + c$ must all be factors of

$$a^4 + b^4 + c^4 - 2b^2c^2 - 2c^2a^2 - 2a^2b^2,$$

which therefore equals the symmetrical expression

$$(a + b + c)(a - b - c)(b - c - a)(c - a \overset{.}{-} b).$$

4E: *A masterpiece by Euler*

The pentagonal numbers form the series

$$1 \quad 5 \quad 12 \quad 22 \quad 35 \quad 51 \quad \cdots \quad ,$$

and we can spot that, in the series that Euler calculated, the indices of every other power of x are the pentagonal numbers. As it happens, however, the connection is even stronger, because if we take the formula for the pentagonal numbers, which is $\frac{1}{2}n(3n - 1)$, then we can calculate pentagonal numbers for negative values of n. When we do so we get the double series

$$\cdots 40 \quad 26 \quad 15 \quad 7 \quad 2 \quad 0 \quad 1 \quad 5 \quad 12 \quad 22 \quad 35 \quad 51 \cdots$$

Euler spotted that the powers in his series use alternate terms from the forward series and the backward series, starting 0, 1, 2, 5, 7, 12, . . .

4F: *Find the factors*

$$1 + x^4 + x^5 = (1 - x + x^3)(1 + x + x^2),$$
$$1 + x^7 + x^8 = (1 + x + x^2)(1 - x + x^3 - x^4 + x^6).$$

4G: *Summing a series*

Call the sum of the series (assuming that it has one) S. Then

$$(1 - x)S = 1 + 2x + 3x^2 + 4x^3 + 5x^4 + \cdots.$$

The effect of multiplying by $1 - x$ is to replace the original coefficients by their differences (apart from the 1 at the start, which does not change.) Therefore, if we multiply by $1 - x$ again, we shall get the series

$$(1 - x)^2 S = 1 + x + x^2 + x^3 + x^4 + \cdots,$$

in which all the coefficients are 1; multiplying for the last time gives

$$(1 - x)^3 S = 1,$$

because the final row of differences consists entirely of zeros. Therefore $S = 1/(1 - x)^3$.

4H: *The three Dutchmen and their wives*

Since each person 'bought as many hogs as they gave shillings for each hog', they each spent a square number of shillings. Moreover, the difference between the square numbers of shillings spent by each husband and wife was the same, 3 guineas, or 63 shillings. Therefore, we need to express 63 as the difference between two squares in three different ways: these are $8^2 - 1^2$, $12^2 - 9^2$, $32^2 - 31^2$. From the remaining information given, Hendrick bought 32 and Catriin 9; and Claas bought 12 and Geertrick 1. So the husband-and-wife pairs are Hendrick–Anna, Claas–Catriin and Cornelius–Geertrick.

4I: *Logic and mathematics*

The engine driver is Brown. Mr R lives in Leeds, and either Mr B or Mr J in London. One third of Mr J's income could not be the guard's. So the guard's neighbour must be Mr B. So Mr J lives in London and is therefore the guard's namesake, i.e. the guard is J. Now B is not the fireman and so must be the engine driver.

4J: *Fermat's Last Theorem*

Suppose that $x^4 + y^4 = z^4$; then we can assume, as in the main text, that x, y and z have no common factor. Then

$$x^4 = z^4 - y^4 = (z^2 + y^2)(z^2 - y^2),$$

and it follows that $z^2 + y^2$ and $z^2 - y^2$ must be separately squares. But this is just what Fermat proved is not possible. Therefore, the original equation is impossible.

4K: *The angle in a semicircle*

Draw a square in a circle, and add one diagonal. We see that one particular angle in a semicircle is a right angle. But all the angles in the same segment will be equal; therefore the angle in a semicircle is always a right angle.

4L: *A special case of Ptolemy's theorem*

Fig. 4.16

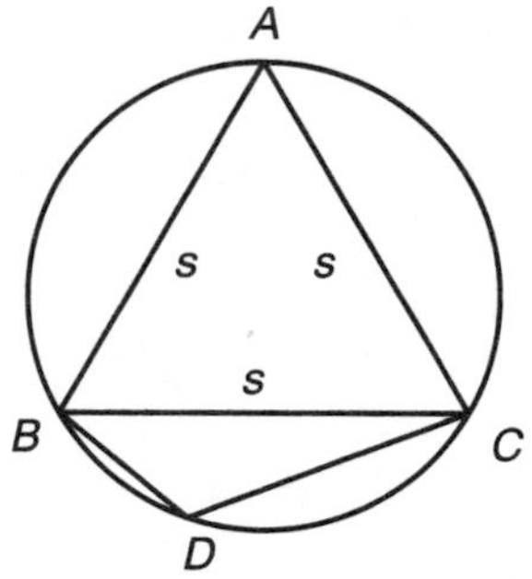

Let s be the side of the equilateral triangle. Then, by Ptolemy's theorem,

$$s \cdot BC + s \cdot CD = s \cdot AC.$$

Therefore, $BC + CD = AC$, whatever the position of C on the arc BD.

4M: *A constant chord*

In Fig. 4.10, join A and Q, as in Fig. 4.17. Then, as T moves round the smaller circle, the angle at T is constant, and so is the angle $A\widehat{Q}B$.

Fig. 4.17

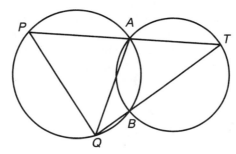

Therefore all the angles of triangle TAQ are fixed, and $\widehat{TAQ}$ is constant. Therefore $\widehat{PAQ} = 180° - \widehat{TAQ}$ is also constant, and so PQ is of constant length.

4N: *Touching circles*

The angles at A, B and C are all constant, and the angle sum of quadrilateral is always 360°. Therefore the angle at which AX and CV meet is also constant. In fact, since

$$A + B + C + (360° - D) = 360°,$$

we see that

$$D = A + B + C.$$

We now join the centres of the four circles (Fig. 4.18), and observe that the angle A is one half of angle P, and similarly B is one half of Q and C is one half of R. Therefore,

$$D = \tfrac{1}{2}(P + Q + R) = \tfrac{1}{2}(360° - S) = 180° - \tfrac{1}{2}S.$$

But the angle at any point on the outer arc of the small circle will be $\tfrac{1}{2}S$, and so D lies on the same circle, on the opposite side of XV. Therefore AX and CV meet on this circle.

Fig. 4.18

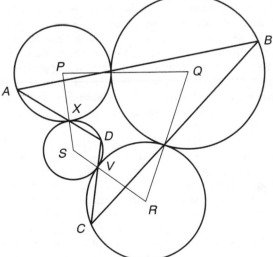

Chapter 5
Creating new mathematical games

However mathematics starts, whether it is in counting and measuring in everyday life, or in puzzles and riddles, or in scientific queries about projectiles, floating bodies, levers and balances, or magnetic lines of force, it eventually becomes detached from its roots and develops a life of its own. It becomes more powerful, because it can be applied not just to the situations in which it originated but to all other comparable situations. It also becomes more abstract, and more game-like. When, finally, after perhaps centuries of 'research and development' a particular portion of mathematics is (so the mathematicians believe!) perfectly well understood, then it has its own rules and procedures, its own objects with their properties and relationships, its own well-known facts and theorems, and it even has its own brilliant tactics and strategical ideas. It is indeed a complete miniature mathematical world.

What happens then? With greater experience, the game will be played better. Results which may have seemed amazing when first discovered will seem familiar and quite clear, even obvious. It will no longer seem so mysterious, or challenging, and so it will not be so attractive to practising mathematicians. More and more types of problem will be solved by standard methods, and so the range of techniques available will increase. Therefore applications will become easier and easier – but the difficult and demanding problems that test the strongest mathematicians may become rare and difficult to find. So mathematicians will have a strong incentive to go outside the game, to ask questions about the game, to develop variations of the game, to try to go beyond its original form.

It is not impossible for mathematicians to go on playing the same game, year after year. This happened to Euclidean geometry towards the end of the last century. Long after most professional mathematicians had moved on to new and more difficult geometrical questions, a small band of devoted enthusiasts continued to explore the legacy of Euclid. They discovered yet more special points, special lines, special circles, which were named, often after their discoverer: the Brocard points were named after Henri Brocard, an officer in the French army and amateur mathematician, who described them in

1875. There were also the Nagel point, the Lemoine points, circles named after Tucker, Neuberg and Fuhrmann, and Kiepert's hyperbola and so on – the list was almost endless. Most mathematicians, however, were not interested. They had long since been forced by their own questions and their own discoveries to leave Euclidean geometry behind.

Pappus of Alexandria (*c.* AD 300) in his *Mathematical Collection* had already proved that, if four concurrent lines cross two lines (see Fig. 5.1), then the ratio of ratios, or *cross ratio* $(AB/BC)/(AD/DC)$, is equal to $(A'B'/B'C')/(A'D'/D'C')$ (Fig. 5.1). This is just one of many propositions that are quite unlike anything in Euclid which clearly developed from the kind of properties that could have been discovered by any carpenter or architect.

Pappus also discovered *Pappus' theorem*, which states that, if three points are marked on each of two lines, and joined as in Fig. 5.2, then

Fig. 5.1

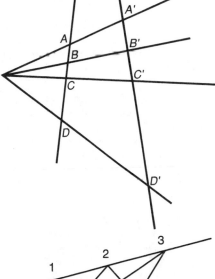

Fig. 5.2

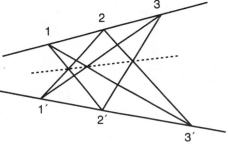

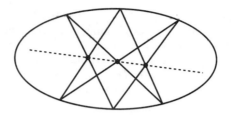

Fig. 5.3

three of the points where the joins meet lie on a straight line (dashed in the figure). If the points are labelled 1, 2, 3, and 1′, 2′, 3′, then the first marked point is the meet of 12′ and 21′, and similarly for the other two marked points. This theorem, like Pappus' cross-ratio theorem, has the remarkable feature that it has nothing to do with either lengths or angles.

In the seventeenth century, Gerard Desargues (1593–*c*. 1662) and Blaise Pascal (1623–62), the philosopher, theologian and mathematician, had also discovered theorems with the same property; these led eventually to the creation in the nineteenth century of *projective geometry*.

Pascal discovered the theorem named after him when he was only 16. It says that if you take any six points on a conic, which could be an ellipse, a parabola or a hyperbola, and join them in the same order as in Pappus' theorem, then the three marked crossings (see Fig. 5.3) lie on a straight line.

This looks much like Pappus' theorem, and indeed the latter is a special case of Pascal's, because a pair of straight lines can be thought of as a 'degenerate' conic. The standard conics can be obtained by slicing a double cone with a plane (Fig. 5.4). If the plane passes through the vertex of the double cone, the slice shows a pair of straight lines.

Fig. 5.4

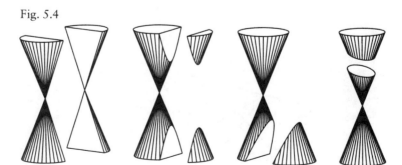

Problem 5A   *The dual of Pappus' theorem*

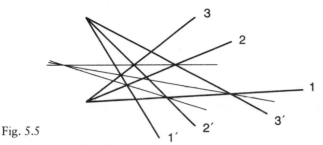

Fig. 5.5

Theorems in projective geometry have what are called *dual theorems* which are obtained by rewriting the statement of the theorem, with lines replaced by points, and points replaced by lines, and with the points where lines intersect switched with the lines through given points. When Pappus' theorem is given this treatment, we get an arrangement like the one in Fig. 5.5. We start with two points, rather than two lines, and three lines through each point, rather than three points on each line. We take the lines in order (they have been marked 1, 2 and 3) and find the six points of intersection, which correspond to the six cross-lines in the original Pappus figure. The first two are the meets of lines 1 and 2' and 2 and 1'. Then the three dashed lines through pairs of these points concur.

Fig. 5.6

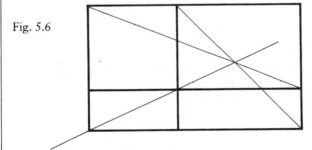

Figure 5.6 is simpler. It is just a rectangle divided into four rectangles by a vertical and horizontal line. As anyone fiddling with the diagram might notice, the three thin lines concur. Are there any other sets of three lines joining the corners of rectangles in this figure which concur? What is the connection between this figure and the converse of Pappus' theorem?

Desargues' theorem concerns concurrent lines and collinear points. The two triangles shown in Fig. 5.7 are in perspective, meaning that lines through corresponding vertices meet at a point, O. Desargues' theorem asserts that, if this is so, then the intersections of corresponding sides lie on a line. They do, on the line marked L.

Desargues' theorem, unlike Pascal's, can be proved rather easily, provided that you are prepared to go into three dimensions to see what it means, and then back into two dimensions. In Fig. 5.8, we are looking at a tripod of lines through O, in three dimensions. This

Fig. 5.7

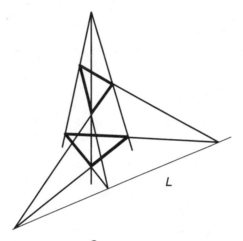

Fig. 5.8

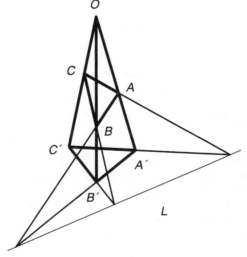

tripod is cut by two planes, represented by the triangles *ABC* and *A'B'C'*. A pair of corresponding sides of these triangles, such as *AB* and *A'B'*, both lie in the plane through *AOB*, so they will definitely meet, unlike two random lines in three dimensions which generally will not meet at all. However, since they also lie in separate planes *ABC* and *A'B'C'*, they must meet on the intersection of these planes, which is the line *L*. Similarly, the meets of *BC* and *B'C'* and *CA* and *C'A'* both lie on *L*. We now *project* the three-dimensional figure onto a plane surface, by imagining a point source of light shining on it and casting a shadow on the plane. The result is a two-dimensional figure in which the theorem is still true. This idea of *projection* gave the new geometry its name. Pascal first proved his theorem for a circle, by traditional methods, and then projected the circle into a conic, and correctly concluded that it would be true for any conic.

Just because projective geometry is so different, because its theorems look different, because it does not deal with lengths and angles, because it uses this new idea of projection, it would be pointless to try to include it in Euclidean geometry, as a sort of extra part or special section. There is no essential contradiction between them, and in theory you can prove every theorem of projective geometry by

---

Problem 5B *Unwrapping Pascal's theorem*

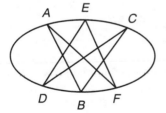 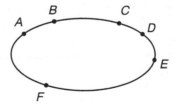

Fig. 5.9

The six points on the conic in the statement of Pascal's theorem can be thought of as the six vertices of a hexagon, *A*, *B*, *C*, *D*, *E* and *F*, in that order, which happens to cross itself three times (Fig. 5.9). The six vertices could also be marked in the normal manner, in sequence round the hexagon, as in the right-hand figure. What will Pascal's theorem say in this case?

Problem 5C   *Brianchon's theorem (the dual of Pascal's)*

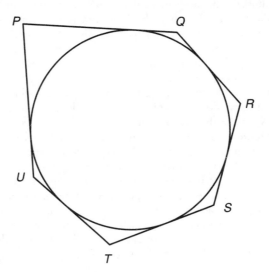

Fig. 5.10

Pascal's theorem also has a dual form. We keep the original conic (in Fig. 5.10 it is a circle) but, instead of taking six points on the circle, we take six lines which touch it, so points are replaced by tangents. Instead of considering the points where opposite sides meet, we consider the lines which join opposite vertices P, Q, R, S, T and U. What can be said about the lines PS, QT and RU?

Euclidean methods. But the results would be horrendously complicated, and the attempt absurd in practice. Much better to treat projective geometry as a new game, a new and beautiful miniature world to explore.

New algebras

The development of new geometries – we shall see some more examples later – was matched by the development of new kinds of algebra, especially the algebra of *complex numbers*, which allows negative numbers to have square roots. In both cases, there had been hints of the new theories in the works of the Greeks. Apollonius (*c.* 262 BC–190 BC) uses ideas which are primitive versions of projection and coordinates: Diophantos met $\sqrt{(-167)}$, the square root of a negative number, when solving a quadratic equation, but did not take it seriously.

The square roots of negative numbers, so-called 'imaginary numbers', turned up again in attempts to solve cubic equations. The first person to publish an algebraic solution of a cubic was Girolamo Cardano (1510–76). (Some cubics had previously been solved by geometrical methods, by the Persian poet and mathematician, Omar Khayyám.)

Cardano was a doctor by profession, who wrote volumes on medicine, astronomy, physics, dreams, chess, music and gambling, and on mathematics. He was the first mathematician to discuss probability, before Pascal and Fermat, and he also discussed what is now called 'Pascal's triangle' (which had been known to Chia H'sien (*c.* AD 1100) in China and may have been known even earlier, to Omar Khayyam).

Cardano obtained the method of solving cubics from Tartaglia, after promising never to reveal the secret. This is ironic, because Cardano was a brilliant algebraist and might well have found such a method himself. Anyway, six years later he discovered that the credit actually belonged not to Tartaglia but to Scipione del Ferro, so decided he was no longer bound by his promise and published the solution in his book *Ars Magna*. This book also showed how to solve equations of the fourth degree by a method discovered by Cardano's pupil and son-in-law, Ludovico Ferrari.

Cardano's method for the cubic was essentially based on the identity

$$(a + b)^3 = 3ab(a + b) + a^3 + b^3.$$

Problem 5D *Cardano on odds at dice*

Cardano discussed probability in his work *The Book of Games of Chance*. Here are three of the problems that he posed:

(a) You throw two dice. What is the chance that you will get at least one 4, 5 or 6?

(b) If you throw the pair of dice three times in succession, what is the chance that you will score a 4, 5 or 6 each time?

(c) What is the chance that, if three dice are thrown, at least one of them will be a six?

His first step was to transform the original cubic equation to get rid of the term in x^2. This left an equation such as

$$x^3 = 6x + 10.$$

Comparing this equation with the identity, when $x = a + b$, we see that they match precisely, provided that

$$6 = 3ab \quad \text{and} \quad 10 = a^3 + b^3.$$

This step is useful, because we now know that $6^3 = 27a^3b^3$, and therefore know the product and the sum of a^3 and b^3, so we can calculate the *difference* of the two cubes, using the formula

$$(p + q)^2 - 4pq = (p - q)^2.$$

When we know their sum and their difference, we can find the numbers a^3 and b^3, and hence find a and b, and so find x.

This process is typical. The problem of solving a cubic is reduced to a quadratic problem, which is reduced to an even simpler problem. The solution of a fourth-degree equation proceeds on similar lines – the task is reduced to that of solving a cubic. (But attempts to solve fifth-degree equations by similar means failed: the new equation was of a higher degree – the sixth – and could usually not be solved.)

Cardano's argument is very elegant, and yet it also leaves some mysteries behind. One mystery is the fact that it apparently leads to just one value of x. In our sample case,

$$3ab = 6, \qquad a^3 + b^3 = 10,$$

we have

$$a^3b^3 = 8.$$

Then, setting $p = a^3$ and $q = b^3$, we get

$$(a^3 - b^3)^2 = 10^2 - 4 \times 8 = 68.$$

So

$$a^3 - b^3 = \pm\sqrt{68}.$$

Taking $a > b$ (the choice $b > a$ gives the same result), we have

$$a^3 = \tfrac{1}{2}(10 + \sqrt{68}) \quad \text{and} \quad b^3 = \tfrac{1}{2}(10 - \sqrt{68}),$$

that is,

$$a^3 = 5 + \sqrt{17} \quad \text{and} \quad b^3 = 5 - \sqrt{17}.$$

Now a calculator gives

$$a = 2.089525 \quad \text{and} \quad b = 0.9571553,$$

and finally $x = 3.0466803$, or $x = 3.046680$ to six significant figures.

A graph comparing the functions x^3 and $6x + 10$, however, shows that they cross at three points – where are the missing solutions?

Cardano found another difficulty when he solved equations such as $x^3 = 15x + 4$, whose solution, by his method, is

$$x = \sqrt[3]{[2 + \sqrt{(-121)}]} + \sqrt[3]{[2 - \sqrt{(-121)}]},$$

which involves the square root of the negative number -121. But the equation has three real roots, including $x = 4$.

Cardano was aware of negative numbers and even considered their square roots. He once proposed the problem 'Divide 10 into two parts such that the product is 40.' He first stated that this was impossible, and then solved the equation anyway, giving the solutions $5 + \sqrt{(-15)}$ and $5 - \sqrt{(-15)}$. He concluded that 'These quantities are "truly sophisticated"' and that to continue working with them would be 'as subtle as it would be useless'.

Cardano, like other early algebraists, was torn between the fact that complex numbers, those 'imaginary' quantities, were apparently absurd, and the fact that they did often work in practice.

It later turned out that not only does every cubic have three roots, but every fourth-degree equation has four roots, and so on. How elegant and simple! How desirable! Once again, mathematicians were forced to accept mathematical objects that they did not, frankly, understand. In this case, a deeper understanding had to wait until the eighteenth century.

Approximate solutions

Cardano was also the first, as it happened, to consider the possibility of finding approximate solutions to equations. This was not a serious practical problem in his day, but it became one in due course, for a variety of reasons. Once it became apparent that equations of the fifth degree and higher could not generally be solved by a formula, it was natural to consider how their roots, which certainly existed, might be found by other means, as approximations.

More generally, as applications of mathematics became more complex, many problems turned up which could not be solved exactly, but for which good practical approximations were essential. Finally, with the development first of desk calculators, and then electronic

Problem 5E   *Using complex numbers*

A complex number consists of a real number plus a multiple of i, the square root of -1. So $3 + 2i$ and $4 - \frac{1}{2}i$ and $-4 + 6i$ are all complex numbers. They are added, subtracted, multiplied and divided by the rules of ordinary arithmetic and elementary algebra, with the difference that whenever i^2 appears it can be replaced by -1. For example,

$$(a + ib)(a - ib) = a^2 - i^2b^2 = a^2 - (-1)b^2 = a^2 + b^2.$$

So, when we are using complex numbers, the sum of two squares can be factorized, as well as the difference of two squares.

Also, by using complex numbers, every quadratic equation has two (possibly coincidental) roots, for example the equation

$$x^2 - 8x + 25 = 0,$$

although the graph of $y = x^2 - 8x + 25$ (Fig. 5.11) clearly shows that it has no real roots. The roots will be two complex numbers of the form $p + qi$ and $p - qi$, whose sum is 8 and product 25. What are they?

Fig. 5.11

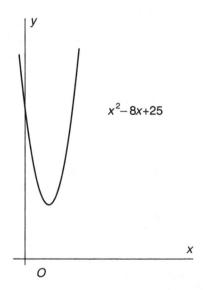

$x^2 - 8x + 25$

calculators and computers, it became practicable to solve complex problems by brute force: it is common today to solve sets of thousands of equations, or to use computer programs which make billions of individual small calculations.

The result of looking for different kinds of solutions is, typically, that different questions are asked, novel ideas are required, new language is invented, and traditional algebra comes to be used in new and striking ways. Let's look at two examples. The next equation has a root between 1 and 2; how can we find it more accurately?

$$x^3 + x^2 = 10.$$

By simple algebra, the equation is equivalent to each of these equations:

$$x^2 = \frac{10}{x + 1}, \qquad x = (10 - x^2)^{1/3}.$$

Let's take the first of our alternative versions of the equation above, and use it to calculate a sequence of approximations, like this:

$x_1 = 2,$ $\qquad$ $x_2^2 = 10/(x_1 + 1),$

$x_2 = 1.82574186,$ $\qquad$ $x_3^2 = 10/(x_2 + 1),$

$x_3 = 1.8811947,$ $\qquad$ $\cdots$ (repeat the same process),

$x_4 = 1.8630035,$ $\qquad$ $\cdots$ and so on.

$x_5 = 1.8689128,$

This process is especially easy to perform on a calculator; but it has the disadvantage that, although it is getting closer and closer to the true answer, it is doing so rather slowly. The second formula also gives a convergent process, somewhat slower than the first.

Checking the last result of the first process, we find that $x_3^3 + x_5^2 = 10.020639$.

Both these approximation processes do work. They do get closer and closer to the real root at each step, yet they raise as many questions as they answer: Why do they work at all? When will they work? (Many such processes actually get *further* from the root at each step.) For practical purposes, an approximate method that converges very slowly to the solution may be useless. Which processes converge fastest? Will a process which is most efficient for *this* equation be most efficient for *that* equation? What will be the error in the solution after N steps? These questions lead to an entirely new branch of mathematics – numerical analysis.

Problem 5F   *Solving quadratics by approximation*

Starting with the equation $x^2 = 10 + x$, we can rewrite it in any of these forms:

$$x = (10 + x)/x, \qquad x = \sqrt{(10 + x)}, \qquad x = \sqrt{(x^2 - 10)}.$$

Each of these transformations can be used to start a sequence of numbers, which might get closer and closer to one of the roots of the quadratic equation. For example, if we pick $x_1 = 3$ as a rough approximation, then from the first transformation we get this sequence:

$$x_1 = 3, \qquad x_2 = (10 + 3)/3 = 4\tfrac{1}{3},$$

$$x_3 = (10 + 4\tfrac{1}{3})/4\tfrac{1}{3} = 3.3076, \qquad x_4 = 4.0233, \qquad x_5 = \cdots.$$

Does this process tend to a root of the original equation? Will either of the other transformations have the same result?

A second method arises from an idea which is now second nature to every mathematician, because it is intimately related to ideas of the calculus, but would have been very strange to Cardano. Suppose that we have an *approximate* solution, call it x_1, which is quite close to the true solution. Then we can write the true solution as $x_1 + e$, where we can be confident that e is small (e stands for the *error*). We know therefore that

$$(x_1 + e)^3 + (x_1 + e)^2 = 10.$$

We now expand:

$$x_1^3 + 3ex_1^2 + 3e^2x_1 + e^3 + x_1^2 + 2ex_1 + e^2 = 10,$$

and then we take a step which is – or rather was, once upon a time – highly original: we just ignore e^3 and e^2 on the grounds that we are assuming that e itself is small and so e^3 and e^2 must be very small indeed. This might seem like cheating, but it would be fairer to say that we are *approximating* the equation, in order to get an approximate solution. We now get the much simpler equation

$$x_1^3 + 3ex_1^2 + x_1^2 + 2ex_1 = 10,$$

from which we calculate that

$$e = \frac{10 - x_1^2 - x_1^3}{3x_1^2 + 2x_1}.$$

This equation for e is not quite accurate, because we *approximated* to get it, but it will hopefully give us a new approximate solution which is more accurate than x_1. Let's check this with the aid of a calculator. The simplest value to take for x_1 is 2, because it is the nearest integer to the solution. Thus,

$$x_1 = 2, \qquad e = -2/16 = -\tfrac{1}{8},$$

and so,

$$x_2 = 1.875.$$

If we substitute this in the original equation, we find that

$$1.875^3 + 1.875^2 = 10.107421,$$

so 1.875 is a much better approximation than 2. However, we can now repeat the same process:

$$x_2 = 1.875, \qquad \text{new } e = -0.00751359.$$

So,

$$x_3 = 1.8674865.$$

Checking once again, $1.8674865^3 + 1.8674865^2 = 10.000375$. This is already very close to 10 for most practical purposes; two steps have given us a much better approximation than four steps of the previous method did.

We mentioned that this method of approximation is connected to ideas in the calculus, which was originally created by Newton (1642–1727) and Leibniz (1646–1716). The calculus gave mathematicians a standard technique for solving problems about gradients of curves, rates of change, and areas under curves, which had previously been solved by many mathematicians, including Fermat and Descartes, but always by idiosyncratic methods which varied from problem to problem.

Here is an example of the triple connection between finding the gradient of a curve, the method of approximation using 'e' which we exploited above, and ideas of errors in measurement. Suppose that we are measuring the movement of some physical object (such as a falling stone) which is proportional to the square of the time, and that our first question is: 'How fast is the quantity changing at time t?' A suitable approximation is $d = 5t^2$, where d is in metres and t is in seconds (see Fig. 5.12). The tricky point about this question is that we are not asking for the average speed over an interval, but the

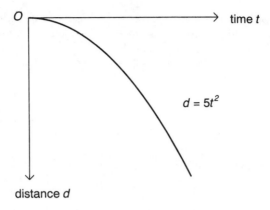

Fig. 5.12

exact speed at an instant, and it is not even clear what the 'exact speed at an instant' means, since the object does not actually move any distance at an instant!

Our solution depends on interpreting the speed 'at an instant' as the limit of an average speed. We first take the position of the object at time t, and the position of the object at a slightly later time $t + e$ (Fig. 5.13). The *average* speed over this interval is the distance moved, which is $5(t + e)^2 - 5t^2$ divided by the time taken, which is e:

$$\frac{5(t + e)^2 - 5t^2}{e} = \frac{10te + 5e^2}{e} = 10t + 5e.$$

Fig. 5.13

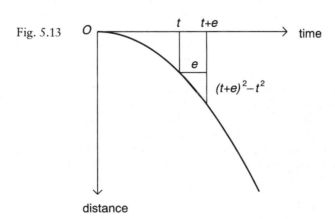

Now we imagine that e gets smaller and smaller, and ask what the limit of this average speed will be? Clearly, it will be $2t$, and this is the speed of the object at time t.

If our object, on the other hand, moved so that the distance travelled after time t was t^3, then by the same reasoning, the average speed from t to $t + e$ would be

$$\frac{(t + e)^3 - t^3}{e} = \frac{3t^2e + 3te^2 + e^3}{e} = 3t^2 + 3te + e^2.$$

As e tends to zero, this expression tends to $3t^2$, which is the rate of change at time t. The connection with our method of approximation is clear enough. The connection with errors in observation is the subject of Problem 5G (see the box).

The discovery of the calculus depended on the prior invention of coordinate geometry, which was created in the seventeenth century by Pierre Fermat and René Descartes. It represents every point on the plane by a pair of numbers, and then translates every statement about points and lines into an algebraic statement about those numbers. So, for example, a circle is represented by an equation which states that the distance of a point on the circle from the centre is constant. If the centre is (2, 3) and the radius is 2 (see Fig. 5.14), then Pythagoras' theorem gives

$$(x - 2)^2 + (y - 3)^2 = 2^2,$$

Fig. 5.14

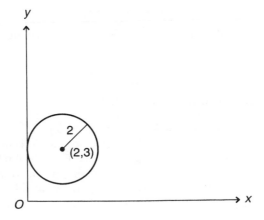

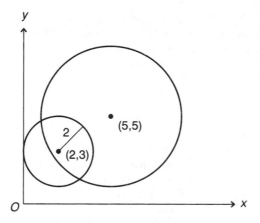

Fig. 5.15

and this is the equation of this particular circle. If we now add another circle (Fig. 5.15), it will also have an equation, in this case $(x - 5)^2 + (y - 5)^2 = 4^2$, and we can calculate the common points of the two circles – the points where they intersect – by finding the common solutions to the two equations. On simplifying, the two equations become

$$x^2 + y^2 - 4x - 6y + 9 = 0,$$
$$x^2 + y^2 - 10x - 10y + 34 = 0.$$

Subtracting, we find immediately that the common points lie on the line

$$6x + 4y - 25 = 0,$$

and we can find the coordinates of the points by finding where this line intersects both circles.

Problem 5G   *How large is the error?*

Suppose that we measure the volume of a cube and we claim that the side of the cube is L and the volume therefore is L^3. Unfortunately, we have made a small error, e, and the actual figures should be $L + e$ and $(L + e)^3$. How large is the error in the measured volume, in proportion to the error in measuring the length of one side?

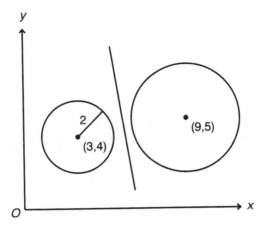

Fig. 5.16

Suppose, however, that we had chosen two different circles, which do not intersect, as in Fig. 5.16. Their equations are

$$(x - 3)^2 + (y - 4)^2 = 4, \qquad (x - 9)^2 + (y - 5)^2 = 9.$$

These two circles have no common points, and yet, when we take the same steps to find these nonexistent common points, by simplifying and subtracting one equation from the other, we discover that 'they' lie on the line

$$6x + y - 38 = 0.$$

This is a real line, which we have marked on the diagram. What does this mean? The secret lies, once again, in the complex numbers. If we allow complex numbers, and therefore complex points, then every pair of distinct nonconcentric circles intersects in two (possibly coincident) points, just as every cubic turns out to have three roots. The points may be both real or both complex, but either way the line through them is real.

As it happens, this real line through the complex points of intersection turns out to be familiar to classical geometers. It is called the *radical axis* of the two circles. Among other properties, it is the line on which lie all of the centres of all the circles which intersect the two given circles at right angles. In this case, we find that complex numbers not only work, but are actually required for us to make sense of the mathematics that we already know.

Problem 5H   *A three-circle theorem*

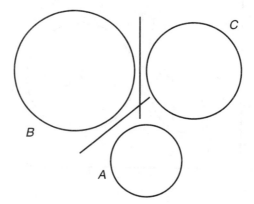

Fig. 5.17

Any point on the *radical axis* of two circles is the centre of a unique circle that cuts both circles at right angles. Figure 5.17 shows three circles, A, B and C, and the radical axes of A and B, and of B and C. What conclusion can be drawn about the radical axis of C and A?

New geometries

In the nineteenth century, many new geometries were discovered. One of the simplest but most powerful was based on the idea that the point half way between two given points is rather like their average: and a point somewhere else on the line joining two points is their weighted average. This is the *barycentric calculus*, introduced by Möbius (1790–1868). In fact it can be regarded as a variant of coordinate geometry, but it is useful enough and powerful enough to be treated separately (though it is a source of confidence that, if necessary, any statement in the barycentric calculus can be translated into a statement in coordinate geometry, so we need have no fear that the rules of barycentric calculus do not work.)

Figure 5.18 shows two points, P and Q, marked on the coordinate plane. The point R, midway between them is indeed their average, if you interpret 'average' to mean averaging the x and y coordinates separately. Leaving coordinates behind, and marking R as $\frac{1}{2}(P + Q)$,

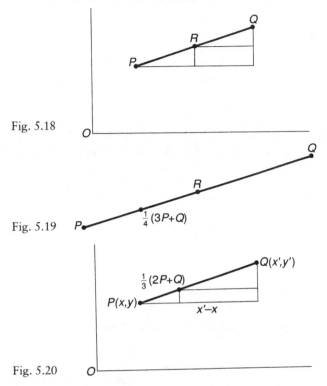

Fig. 5.18

Fig. 5.19

Fig. 5.20

the point S half way between P and R (Fig. 5.19) will be their average, which is

$$\tfrac{1}{2}[P + \tfrac{1}{2}(P + Q)] = \tfrac{1}{4}(3P + Q).$$

Of course, not all geometry can be done by using only midpoints. This expression $\tfrac{1}{4}(3P + Q)$ suggests how we should proceed further. Point S is 3 times as near to P as it is to Q; so, in the expression $\tfrac{1}{4}(3P + Q)$, we can say that P is given a *weight* of 3 and Q a *weight* of 1. The expression, 'weight' suggests a mechanical interpretation: if an actual weight of 3 units were placed at P and of 1 unit at Q, then their centre of gravity would be at $S = \tfrac{1}{4}(3P + Q)$. Following these analogies, the point which is twice as near to P as it is to Q should be $\tfrac{1}{3}(2P + Q)$, and one final check by going back to coordinates confirms that this does indeed make sense: the x coordinate (see Fig. 5.20) is

$$x + \tfrac{1}{3}(x' - x) = \tfrac{2}{3}x + \tfrac{1}{3}x' = \tfrac{1}{3}(2x + x'),$$

and similarly for the y coordinate.

This new geometry, just like Euclidean geometry or projective geometry, has its own strengths and weaknesses. It ought to be very easy and powerful when used to tackle problems about points on given lines, and proportions. In contrast, it says nothing about angles at all, so problems about angles may simply not occur when using this geometry, and angle problems shouldn't be tackled by its aid – use something else.

Let's go back to the centre of gravity of the triangle, which involved, if you recall, the midpoints of the sides. Suppose that, instead of considering the triangle as an area or a flat sheet, we think of its vertices as three separate points. Then the midpoint of each side will be the average of the corresponding pair of vertices. These are easily marked, as in Fig. 5.21. The centre of gravity is where the three medians meet, so it must be some suitable weighted average of A and $\frac{1}{2}(B + C)$, as well as being a weighted mean of B and $\frac{1}{2}(A + C)$ and of C and $\frac{1}{2}(A + B)$.

Fig. 5.21

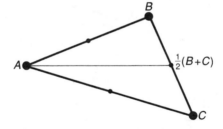

Problem 5I *The centre of gravity of four points*

Fig. 5.22

Figure 5.22 shows four points. The centre of gravity of any pair of them will be the midpoint of the line joining them. Where is the centre of gravity of all four together?

Here we have clues enough to the correct solution, but there is a stronger clue in the fact that the centre of gravity will be an expression which is *symmetrical* in A, B and C. This suggests that it could be $\frac{1}{3}(A + B + C)$. Sure enough,

$$\frac{1}{3}(A + B + C) = \frac{1}{3}[A + 2 \times \frac{1}{2}(B + C)] = \frac{1}{3}[B + 2 \times \frac{1}{2}(C + A)]$$
$$= \frac{1}{3}[C + 2 \times \frac{1}{2}(A + B)].$$

Therefore $\frac{1}{3}(A + B + C)$ does lie on each of these lines, so it must be their common point, and the weights used are always 2 : 1, so it divides each of the medians in the ratio 2 : 1 (Fig. 5.23).

Fig. 5.23

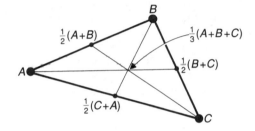

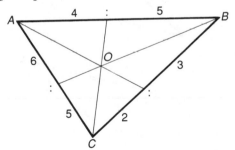

Problem 5J   *A point inside a triangle*

Fig. 5.24

Figure 5.24 depicts a triangle, with a point O marked inside it, and the lines from the vertices through O to the opposite sides, which are divided into parts of the lengths marked. Because all three lines pass through O, these lengths satisfy Ceva's theorem:

$$4/5 \times 3/2 \times 5/6 = 1.$$

How can the position of O be described as a weighted average of the vertices, A, B and C?

It is typical of mathematics that there is an analogy between averages of numbers and centres of gravity, so that it is often helpful to think of one when we are talking about the other – we simply choose whichever way of thinking is simplest at the time. As it happens, there is yet a third way of thinking about the same kinds of problem, which is absolutely essential to mechanics and physics. This uses the idea of a *vector* (see Fig. 5.25).

A vector can be thought of as a length and a direction combined. Velocities and accelerations can be represented by vectors, and so can the simplest movements. These two vectors could represent movements from O to A and from O to B (Fig. 5.26), which we write as $\overrightarrow{OA}$ and $\overrightarrow{OB}$. We can 'add' the two vectors together by making the movements one after the other. It makes no difference whether we make movement $\overrightarrow{OA}$ first and then $\overrightarrow{OB}$, or $\overrightarrow{OB}$ and then $\overrightarrow{OA}$. We still end up at C. We can say that $\overrightarrow{OA} + \overrightarrow{OB} = \overrightarrow{OC}$. Alternatively, we could say that $\overrightarrow{OA} = \overrightarrow{OC} - \overrightarrow{OB}$, or $\overrightarrow{OB} = \overrightarrow{OC} - \overrightarrow{OA}$.

In terms of $\overrightarrow{OA}$ and $\overrightarrow{OB}$ (Fig. 5.27), what would be the movement from O to M, the midpoint of AB? We add the vectors, as before, in either order to get a parallelogram. Then we notice that O and C are

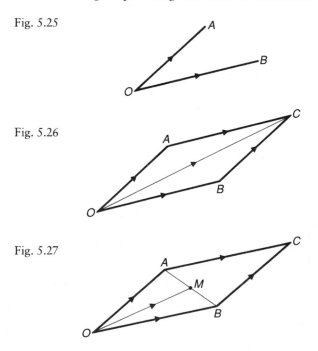

Fig. 5.25

Fig. 5.26

Fig. 5.27

opposite vertices of this parallelogram, of which M is the centre, so the movement from O to M is one half of the movement from O to C. Therefore, $\overrightarrow{OM} = \tfrac{1}{2}\overrightarrow{OC} = \tfrac{1}{2}(\overrightarrow{OA} + \overrightarrow{OB})$. Once again we discover – this time by thinking in terms of straight line movements – that the midpoint of AB is the 'average' of A and B.

All the examples of new mathematical games that we have been looking at developed out of old and familiar ideas which were stretched, extended, added to, or just looked at in a novel manner. There are other sources of mathematical innovation. Some of these lie in science, and we shall see examples in Chapters 9 and 10. Others lie in puzzles and recreations, in posers and teasers that might seem at first glance to be amusing but not serious, suitable occupation for a rainy day but not the stuff of which important mathematics is made. What a mistaken view!

The questions which Diophantos asked about numbers were of no practical importance whatsoever; they were originally puzzling (but difficult) questions which Diophantos enjoyed answering. Yet later, when Fermat studied his copy of Diophantos and asked more and deeper questions, and then when Euler studied Fermat and did the same, followed by others, their researches became the major source of modern number theory.

Another field of modern mathematics, called graph theory (not to be confused with the graphs used in coordinate geometry) has developed explosively in the twentieth century. It grew out of even simpler puzzles, which had been appearing in popular puzzle books long before most mathematicians took them seriously. Here are some examples.

Figure 5.28 gives a plan of the maze at Hampton Court Palace. Early mazes had a religious significance, but the Hampton Court maze was designed purely as entertainment. While walking through

Fig. 5.28

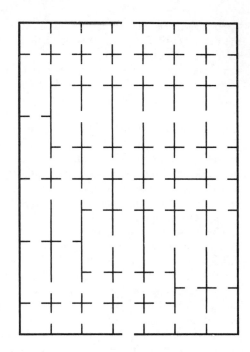

Fig. 5.29

Start at the bottom
entrance and escape
through the top exit, *never
missing* a chance to turn
either right or left.

it, between thick hedges that are too high to oversee, it is easy to get
the feeling that you are lost – though, if the plan of the maze is
simplified by marking all the dead ends and then, as it were, opening
it out, it is seen to be very simple indeed. Certainly anyone who had
the foresight to (say) cross a couple of twigs on the ground at the
entrance to every dead end that they investigated would soon get to
the centre. Modern mazes (like that in Fig. 5.29) are far more
complicated.

Figure 5.30 shows the plan of a house. A burglar, when appre-
hended, claimed that he entered a room by forcing a window at *A*,

Fig. 5.30

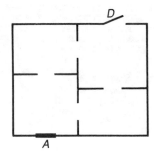

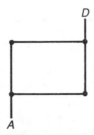

Fig. 5.31

and then worked his way through each room, entering and leaving every room exactly once, and leaving through the back door at D. Why must the burglar be lying? Just as the Hampton Court maze could be represented by a diagram of dots and lines, so can this puzzle. As represented in Fig. 5.31, each room is a dot and each door or entrance is a line. The burglar claims that he started at A and visited each dot once, leaving at D, without going over any line twice.

A similar problem was posed by the Reverend Thomas Kirkman, a distinguished amateur mathematician, in 1855. He asked whether it was possible to find a path on a polyhedron, for example a cube, which visited each vertex once, and only once. Figure 5.32 shows a picture of a cube, as if you were looking close up through one transparent face. This picture naturally consists of lines and vertices.

A solution to Kirkman's problem is today called, sad to relate, a Hamiltonian circuit, after Professor William Rowan Hamilton, who had the same idea just after Kirkman, but who not only happened to be a famous mathematician, but also had the wit to market his puzzle (in 1859) as the Icosian Game. Figure 5.33 shows a view of a regular dodecahedron, seen on the right, viewed as before through one trans-

Fig. 5.32

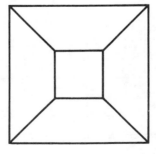

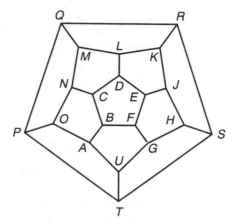

Fig. 5.33

parent face. Hamilton proposed a number of problems with his game, of which this is an example: given the initial sequence B–C–N–O–P, complete a circuit which visits every vertex once only, and returns to B.

The knight's move at chess is difficult for beginners to learn, and lends itself to several puzzles. On the left of Fig. 5.34, the central knight can move to any of the eight marked squares. Of course, if the knight is placed nearer the edges of the board it may have fewer than eight moves open to it. In the corner, it has only two possible moves. In the right-hand figure, the knight starts in one corner of the full chessboard and the puzzle is to make 63 moves, visiting each square once and only once; a harder puzzle is to make 64 moves under the same conditions, returning to the starting square on the final move.

Fig. 5.34

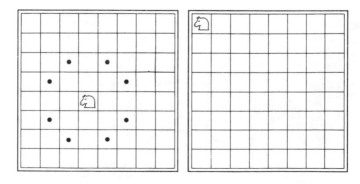

Problem 5K   *A possibility and an impossibility*

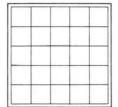

Fig. 5.35

Readers who do not wish to tackle a knight's tour on the full chessboard might like to try the smaller board shown in Fig. 5.35. (a) How can a knight starting at a square of your choice, visit every square once and only once? (b) Why is it impossible on this board for a tour of 25 moves to finish on its starting square?

An additional query: how can knight's tour puzzles be represented by a diagram of dots and lines?

Our last example was proposed by the townsfolk of Königsberg to their most famous citizen, the celebrated Leonhard Euler. The river Pregel runs through the town, surrounding an island and crossed by seven bridges (Fig. 5.36). The puzzle was this: was it possible to take a walk which crossed every bridge once, and only once? Once again, we can simplify the original figure, by replacing the island and the three regions of the town by dots, and the bridges joining them by lines (Fig. 5.37). Trial and error is now easy, and suggests that it is not possible – but why not?

Fig. 5.36

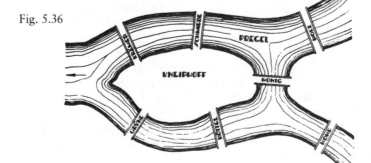

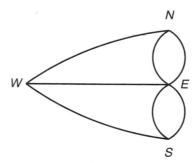

Fig. 5.37

What is the significance of these puzzles? Why do they lead to deep and important mathematics? Ironically, we have already seen one answer to this question, because Euler's relationship between the numbers of vertices, edges and faces of a polyhedron can be thought of as a question about graphs.

Another answer is provided by an important problem in chemistry: in how many ways can given atoms be joined to form molecules? Suppose that we know that a certain chemical has formula C_5H_{12}. This is an example of a *paraffin*: the general formula for a paraffin is C_nH_{2n+2}. The formula means that there are five carbon atoms and twelve hydrogen atoms: each carbon atom has *valency* 4, so it will attach to four other atoms. Each hydrogen atom has valency 1. In how many ways might these carbon and hydrogen atoms be attached? These arrangements of carbon atoms (or carbon and hydrogen atoms), without loops, are examples of what mathematicians call *trees*. Each diagram in Fig. 5.38 is different, and each variant molecule will have slightly different properties. How many different n-carbon paraffins are there? Cayley solved this counting problem in 1875. Today, chemists are able to create extremely complex molecules from simple components, as well as change the form of molecules, for example in genetic engineering, and the problems of the possible structures of complex chemical molecules are more important than ever.

Graph theory is typical of much modern mathematics. Its subject matter is not traditional, and it is not a development from traditional theories. Its applications are not traditional either. Chemistry has always used mathematics, but originally it used arithmetic and algebra to calculate quantities and proportions. Graph theory is not concerned with continuous quantities. It often involves counting, but in integers, not measuring using fractions. Graph theory is an example of *discrete mathematics*. Graphs are put together in pieces, in chunks, rather like Meccano or Lego, or a jigsaw puzzle.

```
    H   H   H   H   H                              H
    |   |   |   |   |                              |
H——C———C———C———C———C——H                    H———C———H
    |   |   |   |   |                              |   H
    H   H   H   H   H                              |   |
                                            H———C———C———C——H
                                                 |   |   |
                                                 H   H   H
                                                     |
                                                 H———C———H
                                                     |
                                                     H
```

```
        H   H   H   H
        |   |   |   |
    H———C———C———C———C——H
        |   |   |   |
        H   H   |   H
            H———C———H
                |
                H
```

Fig. 5.38

Problem 5L   *Trees and paraffin molecules*

Fig. 5.39

Figure 5.39 depicts a tree. Each nonterminal vertex is marked with a large dot, and each terminal vertex with a small dot.

(a) What is the relationship in any tree between the total number of large and small dots and the total number of line segments?

(b) Given that carbon atoms attach to four other atoms, i.e. have a *valency* of 4, and hydrogen atoms attach to only one other (and so have valency 1), why must the general formula for a paraffin with n carbon atoms include $2n + 2$ hydrogen atoms?

(c) How can a paraffin molecule be changed to include a number of oxygen atoms? (Oxygen has a valency of 2.)

Our final example of new mathematical games goes further in the same direction. It starts with Stanislaw Ulam at the Los Alamos laboratories, where he worked on the atomic bomb during the Second World War. After the war, Ulam and some colleagues exploited the primitive electronic computers that were then available, as well as pencil and paper, to investigate what are now called *cellular automata*. The idea is that you start with a grid which could be squares, triangles or hexagons. Some of these *cells* are marked. These are the first-generation cells, from which new cells grow in each generation according to simple rules.

One of Ulam's intentions was to show how very simple rules can create structures which are astonishingly complex. One conclusion that we can then draw is that, when we observe very complicated patterns in nature, such as snowflake crystals observed under a microscope, we need not necessarily imagine that the processes by which they are formed are tremendously complicated – they might be quite simple too. Figure 5.40 illustrates Ulam's very first example. We start with the central square marked 1. The rules are that (1) a new square is added at the next generation to every vacant edge of a square, except that (2) a new square which would have an edge in common with each of two (or more) squares of the old generation is not added. Thus the four empty central squares were not filled at generation 3, because they are each adjacent to two squares of generation 2. If rule (2) is modified so that if two potential added squares would touch even at a corner, at one point, then we add neither of them, then we get patterns such as the one in Fig. 5.41, which has of course more empty space.

Fig. 5.40   Fig. 5.41

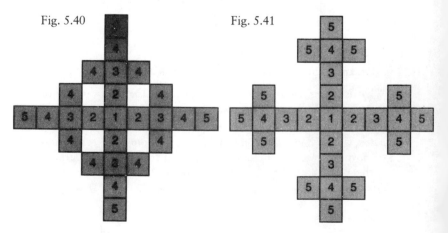

Fig. 5.42

The same rules can be applied on a triangular grid. In fact, if we go back to the just-rigid triangular frameworks which we met in Chapter 1, and start to build outwards from an initial triangle, we get a cellular automaton as illustrated in Fig. 5.42. This framework is always just rigid, because whenever we have a chance to add a triangle which would join a previously placed triangle (and so make it more than just rigid) we reject it.

We can also get 'the same' pattern by starting with hexagons, and making the rules that, at each generation, (1) a hexagon is added to each vacant face, provided that (2) a hexagon which would touch a previously placed hexagon is not placed, and (3) no hexagon is ever placed to complete a row of three adjacent in line. The last rule means that if you have a hexagon sticking out, as on the lower left in Fig. 5.43, then you add new hexagons at the next generation at p and r but not at q.

Fig. 5.43

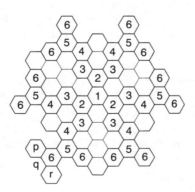

Problem 5M   *A hexagonal cellular automaton*

(a) Why is the hexagonal pattern 'identical' to the pattern on the triangular grid? Why do they match 'exactly'?

   Now for some deeper questions: Fig. 5.44 shows the hexagonal cells numbered, from generation 1 at the centre to generation 11 at the edge.

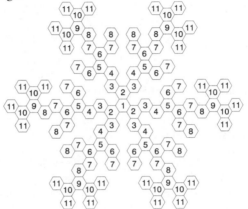

Fig. 5.44

(b) Scientific observation will suggest that the six 'spokes of a wheel', numbered 1–2–3–4– · · · –11 seem to go on for ever in straight lines. If so, why?

(c) The six spokes divide the pattern into six sectors. Alternating sectors are identical to each other, but adjacent sectors are very slightly different. In what does this difference consist?

(d) In the bottom sector of the six, generations 6 and 7 form solid rows, followed by pairs of cells only, one at either end, in generations 8 and 9. The same phenomenon can be seen at generations 2–3 and 4–5. Does this pattern continue for ever? If so, why?

(e) There is a vertical strip of empty cells below cell 1, and two more vertical strips of empty cells are starting below the pair of cells 9. Do these strips go on for ever?

(f) Given all the cells of the nth generation, and no others, is it possible to work backwards and deduce what all the previous generations of cells were?

Fig. 5.45

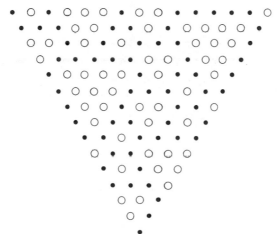

Fig. 5.46

These cellular automata are the simplest kind (in two dimensions), because all that happens is growth according to simple rules. Otherwise nothing ever changes. We can make them even simpler by going down to one dimension, or we can make them more complicated by adding extra rules (or by going up to three dimensions!). Let's look at both possibilities.

In Fig. 5.45, the dots and circles represent a string of cells on an infinite line. The rule is almost the simplest possible: two alike cells give birth to a circle in the next row, which is the next generation. Two unlike cells give birth to a dot. If the initial sequence is chosen at random, then the kind of pattern that results is shown in Fig. 5.46. We enumerate the rows by calling the initial sequence *row 0* (or *the zeroth row*), and then count successive rows 1, 2, . . .

Immediately we see many circle-filled triangles. Even if the initial pattern is random (and perhaps each row is random also), it seems that the pattern as a whole is not. Does that mean that we can *predict* the pattern? A first point to notice is that the state of a symbol in the

Problem 5N *Pascal's triangle, odd and even*

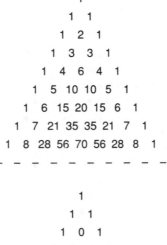

Fig. 5.47

Figure 5.47 (top) shows the first few lines of Pascal's triangle. If we replace all the even numbers by zeros and all the odd numbers by ones (which is the same as replacing each number by its remainder after dividing by 2) then we get the same sort of pattern (bottom figure) as in the one-dimensional cellular automaton illustrated in Fig. 5.46.

(a) Why is this?

(b) This *reduced* Pascal triangle also strongly resembles the hexagonal cellular automaton of Problem 5N. What exactly is the connection between them?

Fig. 5.48

*n*th line depends only on the triangle of cells which is, as it were, resting on it, as in Fig. 5.48. None of the cells to the right or left in the top line has any effect at all.

However, we can go further. In Fig. 5.49, we notice that, if we apply the basic rule to the top left and top right hand corners of the triangle, we can 'deduce' the symbol in the bottom cell, which is the eighth generation (counting the initial row as the zeroth generation), while completely ignoring all the other seven symbols in each top line.

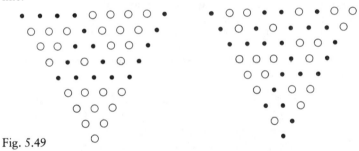

Fig. 5.49

Problem 50 *Deducing row 2^n*

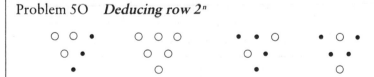

Fig. 5.50

Given a symbol in row 2^n we can deduce its type from the type of the top left and top right symbols in its 'triangle'. The first row to which this applies is clearly row 1 ($n = 0$), which can be calculated directly from the zeroth row (row 0). But we can also immediately calculate the 2nd, 4th, 8th, 16th rows, and so on. Why?

This at once offers the possibility of calculating later rows much more efficiently. If we want to calculate the 22nd row, for example, we can calculate the 16th row directly, calculate the 20th row directly from the 16th, and then calculate the 22nd.

This suggests incidentally that, if the initial (zeroth) row is random (but what exactly does 'random' mean?), then all subsequent rows are random also, because they can be obtained so easily from the first. It is conceivable that a large amount of calculation might somehow turn random rows into less random rows, but it is much less plausible that a small amount of calculation will do so. Therefore the pattern seems to be in all the rows together, not in individual rows.

That is one possibility for calculating ahead. Figure 5.51 shows another. This is an infinite row of circles with a single dot, and the subsequent generations. You could say that in this diagram we see what happens when the perfectly empty row is *disturbed* by a single intruder. How will the effect of the intruder spread like a wave into subsequent rows?

Once again, we get the pattern of odd and even numbers in Pascal's triangle – the reduced Pascal triangle – which is very similar to the pattern of the hexagonal cellular automaton. How does this

Fig. 5.51

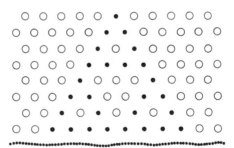

Problem 5P   *The zeros of the reduced Pascal triangle*

Certain rows of the reduced (1 and 0) Pascal triangle (see Fig. 5.47) are all zeros apart from the 1 at each end. How can this be deduced from the fact that row 2^n can be calculated directly from row 0?

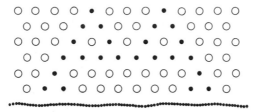

Fig. 5.52

help to calculate the *n*th line? Figure 5.52 shows the combined effects of two such intruders. Where the two waves start to overlap, the combined effect is found by 'adding' the individual symbols, according to the usual rule. We can *superimpose* the results of each separate disturbance: dot on dot = circle on circle = circle, and dot on circle = circle on dot = dot. It follows that, if we want to calculate (say) the type of the marked cell in Fig. 5.53, we could do so by calculating the disturbing effect, *at that cell*, of each of the dot cells in the top row – there are only three of them – and then, superimposing them, calculate the final type.

To do that, however, we need to be able to calculate the type of a chosen entry in the reduced Pascal triangle: see Problem 5T.

This particular one-dimensional cellular automaton is exceptionally simple. This is demonstrated by the very fact that we can predict entries many rows ahead, and even calculate entries by algebraic formulae. These possibilities, in turn, depend on the fact that we can superimpose the effects of one disturbance on another. Even so, most one-dimensional cellular automata are far more complex and unpredictable, as Fig. 5.55 demonstrates.

Fig. 5.53

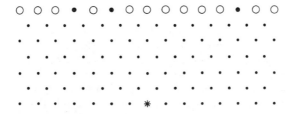

Problem 5Q   *Calculating the reduced Pascal triangle*

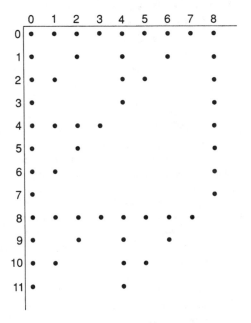

Fig. 5.54

We wish to be able to calculate the parity (odd or even) of any entry in Pascal's triangle, just from its position. Figure 5.54 shows the reduced Pascal triangle (dot for odd, blank for even), swivelled through 45°, with the rows and columns numbered. How can the fact that, for example, the entry in (horizontal) row 11 and (vertical) column 8 is a dot, be calculated from the numbers 8 and 11?

Figure 5.55 shows an example of the Game of Life (*Life* for short) invented by John Conway, who set out to design a cellular automaton which had the greatest richness and interest, for the simplest rules. The game is played on a square grid, each cell of which is either empty (= dead) or occupied (= alive). From a given starting configuration, the rules tell you which cells become dead in the next generation, and which cells become alive, or remain alive. The configuration of four groups of six live cells shown in Fig. 5.55 is an example of

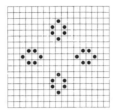

Fig. 5.55

'still life', because according to Conway's rules it never changes. Most configurations are not so simple. They end up either by dying out, or growing for ever or oscillating.

After much experiment, these are the rules Conway came up with.

BIRTH: A cell that's dead at generation t becomes live at generation $t + 1$ if and only if *exactly three* of its eight neighbours were live at generation t.

DEATH by overcrowding: A cell that's live at generation t and has four or more of its eight neighbours live at generation t, will be dead at generation $t + 1$.

DEATH by exposure: a live cell that has only one live neighbour, or none at all, at generation t, will also be dead at generation $t + 1$.

SURVIVAL: A cell that was live at generation t will remain live at generation $t + 1$ if and only if it had just 2 or 3 live neighbours at generation t.

Conway's rules provide a delicate balance between growth, stability and death. Original shapes that differ only slightly will therefore evolve very differently. Figure 5.56 illustrates the life history of a row

Fig. 5.56

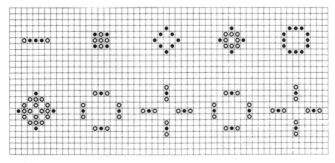

A "line of five" becomes "traffic lights"

of five cells. I have followed Conway's convention of making cells which will survive into the next generation black, and cells which will not survive white.

Figure 5.56 shows a configuration that settles into an oscillation. Other forms oscillate from the start: Fig. 5.57 gives two examples, with Conway's delightful labels. Yet other shapes move through space. In Fig. 5.58 we see a glider, and a lightweight spaceship (there are many spaceships, some of which require smaller spaceships to escort them).

In these figures we are merely glancing at a few of the amazing objects and species which inhabit the *Life* universe. With pencil and paper and patience or with a computer – *Life* might not have evolved without electronic assistance – a strange world of traffic lights, honey farms, glider guns, glider generators, spaceships, eaters, viruses and so on appears.

How mathematical is the Game of Life? If we mean 'How does it resemble traditional mathematics?', the answer is 'Hardly at all!' But, of course, new mathematics is not obliged to resemble the old. The Game of Life, like the one-dimensional automata, is just as mathematical in its basic rules as elementary arithmetic or geometry. Moreover,

Fig. 5.57

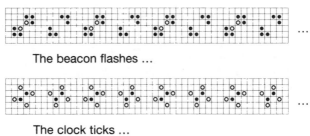

The beacon flashes ...

The clock ticks ...

Fig. 5.58

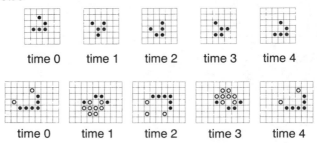

| time 0 | time 1 | time 2 | time 3 | time 4 |

| time 0 | time 1 | time 2 | time 3 | time 4 |

we can explore it like scientists, make inferences, make game-like deductions and use insight and imagination. Indeed, scientific study and insight and imagination are especially needed just because the *Life* universe is so complicated. *Life* is actually so rich and complex that it can be used, suitably organized, as a computer – a suitably large Game of Life could be used, believe it or not, to calculate your tax returns! How remarkable are the games of mathematics!

Solutions to problems, Chapter 5

5A: *The dual of Pappus' theorem*

Take the diagram for the dual of Pappus' theorem, and imagine one of the points we started with disappearing to infinity; then the three lines through it will become parallel. If both original points disappear to infinity in two directions at right angles (see Fig. 5.59), then we will get two sets of three parallel lines, at right angles to each other. This is just what we have in Fig. 5.5. (It would also work if we had a parallelogram rather than a rectangle.) By labelling the two sets of lines, and following the description in the text, we find that there are six ways in which three lines can be chosen to concur. The simplest way to mark these in practice is to pick three of the nine points of intersection so that no two of the points we pick are on the same line, vertically or horizontally (Fig. 5.60). These three points, taken in

Fig. 5.59

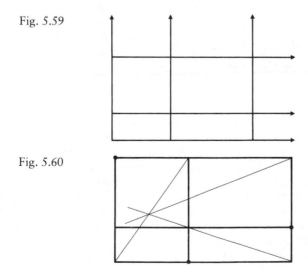

Fig. 5.60

pairs, are then the opposite vertices of three rectangles. The concurrent lines are the three diagonals of these rectangles that do *not* pass through the chosen points. Since there are three ways to choose a vertex in the top line, and then two choices for the vertex in the middle line, after which the bottom choice is fixed in advance, we have $3 \times 2 \times 1 = 6$ choices as expected.

How many of the six points of concurrency will lie inside the rectangle?

5B: *Unwrapping Pascal's theorem*

When the vertices are marked in sequence, Pascal's theorem states that the three meets of pairs of opposite sides lie on a straight line: see Fig. 5.61.

Fig. 5.61

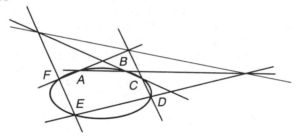

5C: *Brianchon's theorem (the dual of Pascal's)*

The lines PS, QT and RU concur.

5D: *Cardano on odds at dice*

(a) The chance is the same whether you consider one die thrown twice, or two dice each thrown once. The important point is that in either case, the throw of one die has no effect at all on the throw of any other die, or on a throw of the same die again.

There is a 50% or 1/2 chance that you will *not* get 4, 5 or 6 with the first throw, and the same chance that you will not get any of them on the second throw. Therefore the chance that you will get none of them on either throw is $1/2 \times 1/2 = 1/4$. The chance that you *will* get at least one of them, on at least one throw, is therefore $1 - 1/4 = 3/4$.

(b) The chance is 1/2 each time, and so the chance that you will score 4, 5 or 6 on the first *and* the second *and* third throw is $1/2 \times 1/2 \times 1/2 = 1/8$.

(c) The chance of not getting a six on one throw is 5/6. The chance that you will not get a six on any of three successive throws is therefore $5/6 \times 5/6 \times 5/6 = 125/216$, and the chance that you *will* get at least one six is $1 - 125/216 = 91/216$, or about 42%.

5E: *Using complex numbers*

Since $(p - iq) + (p + iq) = 8$, we must have

$$p = 4.$$

So

$$(p - iq)(p + iq) = p^2 + q^2 = 16 + q^2 = 25,$$

and hence $q = 3$. The roots are therefore $4 - 3i$ and $4 + 3i$.

5F: *Solving quadratics by approximation*

By direct calculation, the roots of $x^2 = 10 + x$ are $\frac{1}{2} \pm \frac{1}{2}\sqrt{41}$, or 3.7015621 and -2.7015621. Therefore, if we start with 3 as an approximation, we might expect to get closer and closer to the positive root.

The repeated process $x_{n+1} = (10 + x_n)/x_n$ produces this sequence of approximations:

$$3, \quad 4.333 \cdots, \quad 3.3076 \cdots, \quad 4.0232 \cdots,$$
$$3.4855 \cdots, \quad 3.8689 \cdots, \quad 3.5846 \cdots, \quad \cdots .$$

This sequence is tending to the root, but it is doing so rather slowly, with alternate terms being larger than and less than the root. The second process in the problem also tends to the root, but much faster. By 8-digit calculator:

$$3, \quad 3.6055 \cdots, \quad 3.6888 \cdots, \quad 3.6998 \cdots,$$
$$3.7013 \cdots, \quad 3.70153 \cdots, \quad 3.7015577 \cdots, \quad 3.7015615 \cdots,$$
$$3.7015620 \cdots, \quad 3.7015621 \cdots, \quad 3.7015621 \cdots .$$

In contrast, the third process leads at once to $\sqrt{-1}$ if we start with 3 as an approximation; if we take 4 instead, then it leads to the square root of a negative number after two steps.

These differences illustrate the variety and subtlety of such approximations. When, why, and how they work, and how efficient they are, are all tricky questions.

5G: *How large is the error?*

The volume of the cube, according to our measurement, is

$$(L + e)^3 = L^3 + 3L^2e + 3Le^2 + e^3.$$

If the error e is only small, then we can ignore e^2 and e^3, to get an approximate answer, and the error is $3L^2e$, and (error)$/e = 3L^2$, which is proportional to the surface area of the cube.

5H: *A three-circle theorem*

The third radical axis passes through the meet of the other two. In other words, all three concur. If the circles do not meet, then, as indicated in the text, they still have radical axes when taken in pairs, and these will still concur. The point where they concur is the unique centre of a real circle which cuts all three at right angles (Fig. 5.62).

Fig. 5.62

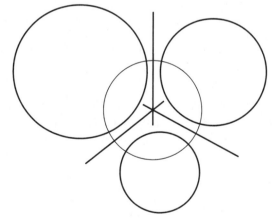

5I: *The centre of gravity of four points*

The centres of gravity of the points taken in pairs, P–Q and R–S, are the opposite midpoints M and N. The centre of gravity of all four points is therefore the midpoint of MN (see Fig. 5.63). Since the points could be split into pairs in two other ways, as P–S and Q–R, or P–R and Q–S, we can find two other pairs of midpoints, whose

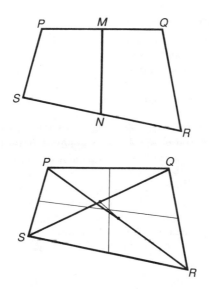

Fig. 5.63

midpoints are also the centre of gravity. This fact can be checked by drawing, as in the lower figure.

5J: *A point inside a triangle*

We can start by imagining weights of 5 and 4 respectively at A and B (Fig. 5.64). Then, to get the correct weighted average for the side AC, we place 6 at C, and this also gives us the correct weighted average between B and C.

Fig. 5.64

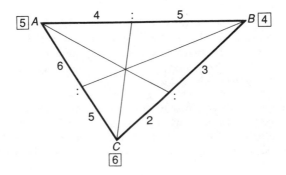

5K: *A possibility and an impossibility*

A pretty solution starts in one corner and hugs the edges for as long as possible. This might seem an arbitrary strategy, but in many knight's-tour problems it does work in practice. The tour shown in Fig. 5.65 (left) ends up at the centre cell. A tour on a board with an odd number of cells cannot return to its starting cell. The simplest way to see this is to colour the board, as on the right, and then consider that a knight always moves from a black to a white square, or *vice versa*. It follows that a complete tour by a knight will always take an even number of moves. The smallest rectangular board on which a knight's tour is possible is 3 × 4, and Fig. 5.66 shows one of the possible tours. The twelve cells can be represented as a network as in Fig. 5.67. Only four cells, namely 7, 4, 6 and 9, can each be reached from three other cells. Although the 3 × 4 rectangle is so small, it is still far easier to spot the possible knight's tours – and to see why almost successful attempts fail – on this graph than it is on the actual board.

Fig. 5.65

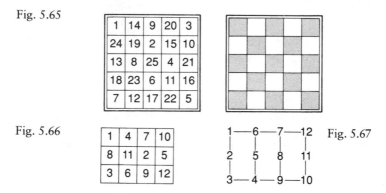

Fig. 5.66

Fig. 5.67

5L: *Trees and paraffin molecules*

(a) The total number of dots, of both kinds together, is one more than the total number of line segments. The easiest way to see this is to consider that every time you chop off a 'branch', you lose one dot and one line segment, until finally you are left with one line segment, with a dot at each end.

(b) A carbon atom attached to four hydrogen atoms could not be attached to any other carbons, and could not be a part of a larger molecule. Therefore, the carbon atoms must be attached to each other. When two carbon atoms are joined, two links will be used up, leaving each with a valency of three. If there are n of them, there will be $n - 1$ pairs using up $2n - 2$ links, out of a total of $4n$ links, leaving $2n + 2$ for the hydrogen atoms.

(c) A single oxygen atom with a valency of 2 can only be placed between a carbon and a hydrogen atom or between two carbon atoms. If there are n carbons, that makes $n - 1$ positions for one oxygen between the carbons, and $2n + 2$ positions where hydrogen atoms are attached, a total of $3n + 1$. That is just the start of an answer, because when more oxygen atoms are added, they can also go between other oxygen atoms ... and the situation is becoming very complicated!

5M: *A hexagonal cellular automaton*

(a) Because no three hexagonal cells can be adjacent, or in a straight line, one third of all the cells in the grid can never be occupied. These are shaded in Fig. 5.68. If we now join the centres of these forbidden cells, then the hexagonal automaton is divided into triangles, and there is an exact match between these triangles and the triangles of the triangular automaton.

Fig. 5.68

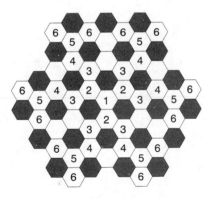

(b) The cells in one generation all lie inside a regular hexagon with cells of that generation at its vertices. Therefore there can be no cells outside this hexagon to inhibit the cells at the vertices from growing in the next generation, when the whole statement just made will once again be true: so the spokes grow for ever.

(c) Three of the sectors grow outwards from cells marked 2–1–2, and the three alternate sectors grow identically, but from cells marked 3–2–3. So the patterns are the same; but, at corresponding points, the cells in one sector are a generation ahead of (or behind) the cells in the adjacent sectors.

(d) Yes, it does. To see this, simplify the pattern by deleting all the cells that did not grow outwards, away from the centre, when they were formed (Fig. 5.69). The pattern is now composed of triangles of cells. Suppose that the cells form a complete row at generation 7. Then the single cells growing at each end at generations 8 and 9 will develop into triangles of cells identical to those developing from cells 4 and 5, and will eventually form two rows, identical to row 7, meeting in the middle, and forming the complete row 15. Similarly, the ends of row 15 lead to two triangles of cells which will eventually form two duplicate rows like 15, to make row 31. And so on.

Fig. 5.69

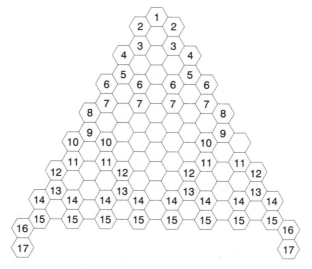

(e) The central strip of empty cells does go on for ever, but the rows starting below 9 will soon be cut off by triangles of cells growing on either side.

(f) No, it is not, curiously enough. In Fig. 5.70, given the 5th generation, there are two sets of positions where the 4th generation might be located. However, the 5th and 4th generations together do fix all the previous generations.

Fig. 5.70

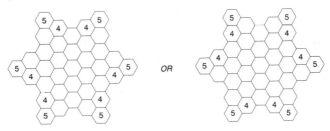

5N: *Pascal's triangle, odd and even*

(a) The same pattern occurs because even and odd numbers obey the rule that 'even + even' and 'odd + odd' are both even, but 'odd + even' and 'even + odd' are both odd, exactly matching the rules for circles and dots.

(b) If we take one sector of the hexagonal automaton, then we can see that, with the exception of the first cell, its cells exist in pairs: a cell from generation $2n - 1$ and the adjacent cell from generation $2n$; see Fig. 5.71. If we think of these pairs as single cells, then there is a perfect correspondence between these double cells

Fig. 5.71

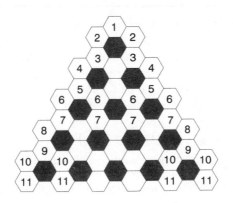

and the single cells of the reduced Pascal triangle, as long as we include only pairs of cells that are 'pointing away', as it were, from the centre (so that parts of the pattern have to be deleted, leaving triangular spaces which match the empty triangles in the reduced Pascal pattern). Thus the four double cells $\binom{5}{5}\binom{5}{5}\binom{5}{5}\binom{5}{5}$ correspond to the row 1111, which is row 3 of the reduced Pascal triangle.

5O: *Deducing row 2^n*

Suppose that the bottom symbol of a triangle can be calculated from the top left and top right symbols, by the usual rules. Then if we fit together a pair of such triangles (Fig. 5.72) we can bypass the middle row to calculate the bottom symbol from the topmost symbols.

So, from two triangles with side 3, we can, by overlapping them, use one triangle of side 5, and by overlapping two of side 5, we can use one of side 9, and so on. In general: $2(2^n + 1) - 1 = 2^{n+1} + 1$ (the subtraction of 1 is to account for the overlap). Since we are counting rows from zero, the bottom of the triangle of side $2^n + 1$ is in row 2^n.

Fig. 5.72

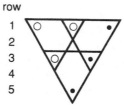

5P: *The zeros of the reduced Pascal triangle*

Row 2^n of the reduced Pascal triangle will depend on the top-left and top-right symbols in the initial (zeroth) row, and these will always be both circles except for two pairs, the single dot constituting the left-hand symbol of the right-hand pair, and the right-hand symbol of the left-hand pair, so only those two pairs will produce dots in row 2^n.

5Q: *Calculating the reduced Pascal triangle*

The rule is this: express the two numbers in base 2, that is, as sums of powers of 2: thus, $8 = 8$ and $13 = 8 + 4 + 1$. If the same power of 2 appears in both representations, then that cell will be blank.

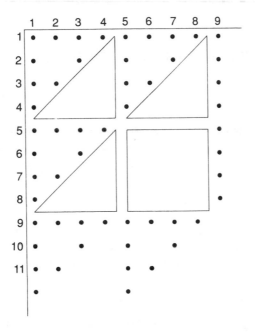

Fig. 5.73

To see why this is so, take a triangle of blanks, and divide it into a square and two smaller triangles, as in Fig. 5.73. If the cell we are considering lies in the square, then both its coordinates have their largest powers of 2 in common. Suppose that it lies in one of the triangles. Take the triangle of the same size above or to the left which is nearest to the corner, and in this triangle take the cell corresponding to our initial cell, in the corresponding position. It has the same representation in powers of 2, except that the largest power is missing.

Once again, consider whether it lies in the square of one of the two triangular wings. If it lies in the square, then (ignoring the largest power of 2, which has been lost) the largest powers of 2 remaining match in both coordinates. If, however, it still lies in a triangular wing – move down to the matching triangle that is nearer to the corner, and repeat the argument.

Eventually, it must lie in the square, and have a matching power of 2 in its representation, because the smallest 'triangle' is just a single cell.

Chapter 6
Perception and imagination

'The moving power of mathematical invention is not reasoning but imagination.'

Augustus de Morgan

How often have you heard someone say, 'I see what you mean,' or 'I don't see what you're getting at'? In modern English the verb 'see' is used more often to refer to understanding than to sight. Vision in mathematics stretches all the way from looking straight at something in front of your eyes, to metaphorically 'seeing' an idea that is totally non-visual, and the latter kind of 'seeing' is just as common and just as important as the former.

Sylvester thought that mathematics was essentially about seeing 'differences in similarity, similarity in difference'. Mathematicians 'see' connections and relationships, but they also see subtle distinctions, and they see these just as readily in arithmetic and algebra as

---

Problem 6A   *Traditional riddles*

The man who made it did not want it, the man who bought it did not use it, and he who used it never saw it.

Give it food, it will live. Give it water, it will die.

You can only hold me for a while, you lose me sometimes, yet I am with you all your life.

What belongs to you, which you would not get rid of, but other people use it more than you do?

What is broken as soon as you name it?

they do in geometry or graph theory. Mathematics in this respect much resembles riddles, and requires the same kind of mentality – which fortunately everyone seems to possess. Our Victorian ancestors were willing to take seriously the kinds of puzzle that today only appear in children's comics and Christmas crackers.

The very forms of traditional riddles, which are conventional and recognized immediately, are reminiscent of the kinds of question that mathematicians ask: why is a . . . like a . . .? What do you get if you cross a . . . with a . . .? What is the difference between a . . . and a . . .? I am . . . and . . . and . . . What am I? Because riddles are so conventional, the kind of answer expected is known in advance. To the riddle 'What is the difference between a hill and a pill?' the response, 'One is smaller than the other,' is not acceptable; the correct answer, recognizable by its twist, is 'One is hard to get up, the other is hard to get down.' (Is there perhaps another answer with as good a twist?)

Riddles can even involve the logical feature of self-reference: to the question, 'What is the difference between an elephant and a pillar box?' the expected response is, 'I don't know, I give up,' to which the riddler responds, 'Then I won't send you to post a letter!' This riddle refers to the riddle form itself – it relies on the usual expectations and then confounds them.

In mathematics, the same kinds of assumption are made: the mathematician is not interested in any old feature, but in significant features, features with a twist. It is true that one difference between a cube and an octahedron (Fig. 6.1) is that one has triangular faces and the other's are square, but this is rather obvious. The deeper and more satisfying difference is that (while they each have twelve edges) the cube has six faces and eight vertices, while the octahedron has eight faces and six vertices. This riddling response is even more significant when you realize that it is not a coincidence of counting, but a consequence of the deeper geometrical relationship between

Fig. 6.1

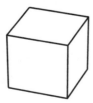

the two solids, which can be expressed, for example, like this: you can inscribe an octahedron in a cube (and *vice versa*) by marking one vertex of the octahedron at the centre of each face of the cube.

We have already seen many examples of striking perceptions. Lhuilier's perception of polyhedra that did not obey Euler's relationship, which was prompted by examining actual mineral crystals, illustrates the value of practical experience in stimulating the geometrical imagination. The proof of Desargues' theorem, by going into three dimensions (p. 120) is another imaginative move, as is the analogy between midpoints and centres of gravity (p. 135).

Cardan's matching of an algebraic identity against a cubic equation (p. 123) is a very important kind of 'seeing' in mathematics. So is the perception by Oresme of an elegant way of grouping a series (p. 87) and Ramanujan's ability to spot fruitful transformations of series (p. 90) and the observation by Auriefeuille and Lucas of the identities that Landry had missed (p. 93). The different ways of looking at the pivot' theorem (p. 109) are also typical, and the different ways of looking at the successive differences of a series (p. 32).

Mathematics depends on the ability to 'see', literally and abstractly, in novel and unexpected ways, which is another way of saying that it depends on the brilliance and subtlety of the human brain which just happens to be wonderfully adapted to this very purpose. We are good, it is true, at straight calculation, but most of us are neither that fast nor that accurate, and computers can beat us on both scores; however, when it comes to perception, to the insights that guide and

---

Problem 6B   *Mathematical riddles*

What is the largest number less than 1?

We are all different, there are an endless number of us, and yet we are all equal to each other.

What is the centre of a triangle?

I multiply without making larger or smaller. What am I?

When multiplied by myself I increase by no more than when I am added to myself. What am I?

What do you get if you cross a cube and an octahedron?

Constantly turning, I never return to my starting point. What am I?

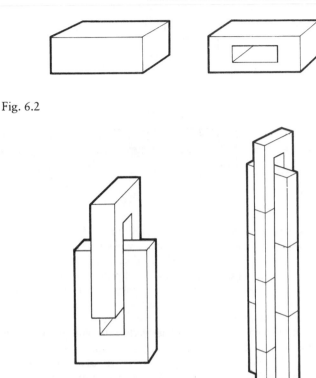

Fig. 6.2

Fig. 6.3

direct calculation, we are wonderful! And computers, at the end of the twentieth century, are nowhere!

Even the apparently simplest and most visual problems can require twists of imagination. Copies of the brick illustrated in Fig. 6.2 (left) can be used to fill space, leaving no gaps, simply by piling them up in layers in the usual way. Suppose however that a hole is cut in the brick, as in the right-hand figure, rather like one of Lhuilier's polyhedra: might it still be possible to fill space with copies of the brick? The tempting response is 'No, of course not!' However, it depends on the shape and size of the hole. If the hole is just the right size, as in Fig. 6.3 (left), then the bricks will link in a chain, to make a column with a cross-section that is a Greek cross – as in Fig. 6.3 (right).

Fig. 6.4

Fortunately, Greek crosses tessellate the plane (Fig. 6.4), so that many of these chains of bricks will indeed fill space without gaps.

John Edson Sweet was an engineer who (like many engineers) had a very good three-dimensional imagination. He was shown a diagram like that in Fig. 6.5, which shows the three pairs of external tangents of three circles of different sizes. It appears that the three points where the tangents meet, A, B and C, lie on a straight line.

Sweet's reaction was, 'Yes, that is perfectly self-evident.' He then explained that he visualized the figure as showing three spheres lying on a flat surface. The pairs of tangents were circular cones which touched the spheres in pairs. As well as the table top that they are resting on, there is another plane which can be laid across the spheres, to touch all of them, and the cones. The vertices of the cones must lie on the line in which these two planes meet, so they, of course, lie in a straight line. This visual argument, which is similar to

Fig. 6.5

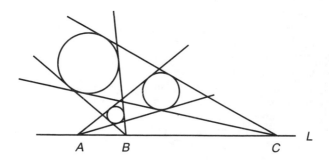

Problem 6C   *Slicing a cube*

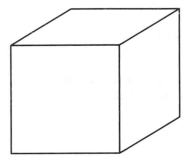

Fig. 6.6

Fig. 6.6 shows a cube. How can it be sliced with a flat knife to display a cross-section which is a regular hexagon? Now suppose that many cubes are stacked in the usual way to fill space. How can this mass of individual cubes be sliced to show new hexagonal faces, in such a way that a new space-filling polyhedron is created?

the argument for Desargues' theorem, is perfectly sound. If you take a plane slice through the centres of all three spheres, you recover the original figure.

There is another way to think of this figure (see Fig. 6.7). Let the distances of the centres of the circles X, Y, Z from the line L be x, y, z respectively. Imagine that you start with the smallest circle, X, and from point A you enlarge it into circle Y. The ratio of this enlargement is the ratio of the distances of the centres from *any* line through A,

Fig. 6.7

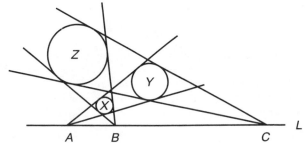

such as the line L, so it is equal to y/x. Now enlarge circle Y, from C into circle Z. The ratio of this enlargement is z/y. Now comparing Z and X, the circle Z is an enlargement of X with ratio $z/y \times y/x = z/x$. Therefore Z is an enlargement of X from a point on the line L, that is, from the original point B. So B lies on L.

It is one of the marvels of mathematics that everything can be seen in different ways from different points of view, sometimes literally, sometimes metaphorically speaking. This is indeed a major reason for the tremendous power of mathematics! The reason why fractions such as $\frac{2}{3}$ are so useful in everyday life and in mathematics is basically that $\frac{2}{3}$ *can mean* two thirds added together; but it can also mean the result of dividing 2 into 3 parts; or it can also mean the answer to the question, 'How many times does 3 go into 2?'

Unfortunately, this also explains why school pupils have such difficulties with fractions. Wouldn't it be simpler if only each fraction had one meaning and occurred in one kind of situation only? Of course, but what then would be the use of fractions! Children also have difficulties with algebra, for the same reason. Wouldn't it be simpler if x was always an *unknown* quantity – that is, a determined but hitherto uncalculated quantity – but, no, it is sometimes an unknown and sometimes a variable, and sometimes it switches from one to the other while you aren't looking. How confusing! But how flexible and powerful!

In time, their teacher hopes, they will clamber over this hurdle and start to learn to play the game of algebra fluently. They will then appreciate the symmetry which is hidden in an ordinary quadratic equation, such as:

$$T^2 + 20 = 7T.$$

On the face of it, this does not look at all symmetrical. Should it be? Well, the graph of T^2, or of any quadratic expression, is a parabola, which is obviously symmetrical. Should not the algebra reflect this symmetry? It should and it does. Divide throughout by T:

$$T + 20/T = 7.$$

Provided that we look rightly, we will see the symmetry. We have to 'see' that this equation as saying: 'Here are two numbers, labelled T and $20/T$, whose product is 20 and whose sum is 7.' Put like that, the symmetry is much clearer. If we call those two numbers, say, p and q, it is clearer still. The equation says that, if $T = p$ and $20/T = q$, then

$$pq = 20 \quad \text{and} \quad p + q = 7.$$

The symmetry is complete. As it happens, the same symmetrical result could have been achieved by a different route. We start by writing the original equation as

$$20 = 7T - T^2 \quad \text{or} \quad 20 = T(7 - T).$$

The equation 'says' the same as before: here are two numbers, this time labelled T and $7 - T$, whose sum is 7 and whose product is 20.

*

Problem 6D   *The symmetry of the cubic*

The quadratic equation $x^2 - 7x + 12 = 0$ has two roots, 3 and 4, and it is true that $7 = 3 + 4$ and $12 = 3 \times 4$. The following cubic equation has three real roots, 2, 3 and 4:

$$x^3 - 9x^2 + 26x - 24 = 0.$$

Bearing in mind possible analogies with quadratic equations, how can the relationships between the roots and the coefficients, -9, 26 and -24, be expressed symmetrically?

Problem 6E   *The smallest triangle*

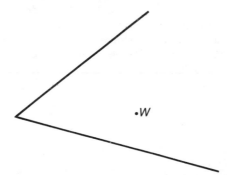

Fig. 6.8

A farmer wishes to fence off the smallest triangular area possible by using one straight fence to join the straight hedges shown in Fig. 6.8. However, the fence must pass through the point W which represents a water-trough that must be available to animals on both sides of the fence. How should the farmer place the fence?

Most people think that mathematics is highly abstract. In a sense this is undoubtedly true, and most mathematicians would agree. Yet mathematicians themselves are continually going back to everyday modes of thinking in order to help them to do mathematics better. They are especially likely to think in terms of movement and change. Although mathematicians have long since ceased to think of an infinite series as 'moving' closer and closer to its limit, like a tortoise just creeping up to the finishing line, they often use such dynamic metaphors to describe what they are doing.

Who hasn't thought of a sequence of integers as one object which actually does things like increasing slowly, or making sudden jumps? We have already referred to the 'prime number race' (p. 55) as if two sets of prime numbers were in a competition for a prize. The mathematician Mark Kac (1914–84), talking about a particular property of the primes, referred to 'the primes playing a game of chance'. The primes do not actually *do* anything themselves, and yet how active they can seem to be! Mark Kac elsewhere quoted from Plutarch, on the Greek use of mechanical devices as a way of thinking about, and solving, geometrical problems. A geometrical figure is, on the face of it, a collection of lines and points, fixed on a plane surface or perhaps fixed in space. Yet geometers continually think of geometrical figures in physical terms.

Figure 6.9 shows a deltoid, a curve first studied in 1745, by Euler. Like several of the curves studied by the Greeks, it is defined by a mechanical motion, and by analysing this motion we can deduce properties of the deltoid. A small circle rolls inside a larger circle, whose diameter is three times that of the smaller. The path of a fixed point on the small circle is a deltoid. Because of the ratio of the diameters, the fixed point will strike the outer circle three times before its path repeats (Fig. 6.10).

Fig. 6.9

Fig. 6.10

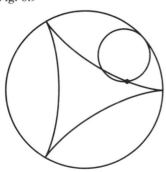
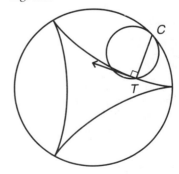

Problem 6F *Cardano's spur-wheel problem*

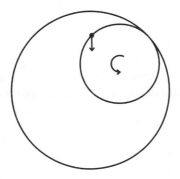

Fig. 6.11

In Fig. 6.11, a circular wheel rolls inside a fixed circle of double its diameter. What is the path of a fixed point on the circumference of the rotating wheel?

Let's start by looking at the tangent to the deltoid at the point T. As the small circle rolls inside the larger (see Fig. 6.10), the point T is instantaneously rotating about the point contact, C. Therefore the tangent at T is perpendicular to TC.

Is there not a kind of symmetry here? If we extend CT to meet the circle again, at D (Fig. 6.12), then T could be instantaneously rotating about D. Might it be rotating on another rolling circle? If we complete the triangles formed by C, T, D and the centre of the two

Fig. 6.12

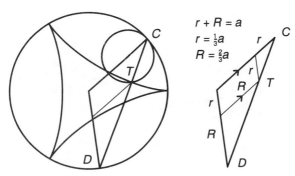

$$r + R = a$$
$$r = \tfrac{1}{3}a$$
$$R = \tfrac{2}{3}a$$

circles, then we can add the thin lines, shown in the right-hand figure, to identify the centre of a larger circle, double the size of the smaller, which is touching the large circle at D and on which the point T is moving.

What will happen a moment later? T will have moved ahead a minuscule amount, but it will still be carried by the small circle, by definition, and also by the medium circle, by symmetry. In other words, the deltoid can also be defined as the path of a fixed point on a circle two-thirds the size of the large circle, also rolling inside it.

This *choice* of an almost-physical way of thinking about the deltoid, as if we had an actual mechanical model on which we could experiment, is an example of perception at the start of a problem. Spot the best way to see the problem, and it may be half-solved before you take another step. Set out with a dud view of the situation, and you may work very hard, but get nowhere. Perception has a different role in spotting the consequences of theorems. A special case is, well, the 'special case' of a theorem. Here is a case in point: Fig. 6.13 was constructed by first drawing the two heavy triangles, which are similar to each other. Then their corresponding vertices were joined, and using these as bases, three different similar triangles were constructed. Their free vertices form a third triangle similar to the original heavy triangles.

Since we can choose any pair of triangles at all to start with, and the three added similar triangles can be any shape at all, this is a very general theorem, and we can expect to get some novel results by taking special cases. Figure 6.14 shows the start of the first. A right-hand vertex of one bold triangle and the left-hand vertex of the other

Fig. 6.13

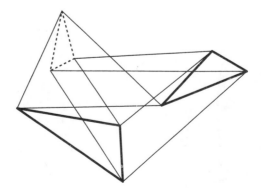

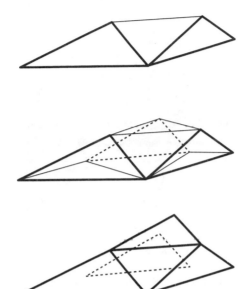

Fig. 6.14

Fig. 6.15

Fig. 6.16

coincide. Two of the lines joining corresponding vertices are now the bases of these triangles, so only one of the three has to be added. We now construct three similar triangles on 'the lines joining corresponding vertices' as bases (Fig. 6.15). We have made these triangles rather small, so that two of the free vertices actually lie inside the original bold triangles. The theorem now says that the free vertices form a triangle similar to the original pair. However, the figure seems a bit unsymmetrical, with two vertices inside the original bold triangle and the other alone at the top, so we will increase the symmetry by adding a third bold triangle, similar to the original pair, as in Fig. 6.16. Now the theorem says that, given three similar triangles, with pairs of common vertices arranged as in Fig. 6.16, and three points, similarly situated, one in each triangle, these points form a fourth triangle of the same shape.

This theorem looks and sounds markedly different from the original theorem, but we can go further, in two directions. The first approach is to set the diagram of the theorem in a wider context, and to realize that similar triangles can actually tessellate; that is, by adding more similar triangles we get a tiling of the whole plane, using two types of triangular tiles, each type of consistent shape but varying in size (see

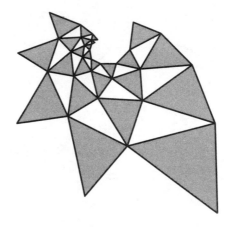

Fig. 6.17

Fig. 6.17). This tiling represents a transformation of the plane in which the entire plane is rotated and at the same time contracted towards the centre (or expanded outwards from the centre, depending on which way round you look at it).

The same transformation can be thought of as sliding a triangle about the plane, always keeping its shape the same, but varying its size and orientation. This is, of course, an unusual way of looking at the usual Euclidean plane, but it is also an advantageous viewpoint from the point of view of the present theorem because, if we have this 'spiral similarity' clearly in mind, then we can instantly *perceive* that Fig. 6.16 is a small section of this spiral tessellation, and that the theorem is therefore obviously true.

The second approach is to specialize the diagram, to get a theorem that is actually quite well known. We do so by making the original bold triangles equilateral, and instead of thinking of them as sharing pairs of vertices, we 'see' them as constructed on the outside of the edges of an arbitrary triangle. For the similarly situated points inside each equilateral triangle, we choose their centres (see Fig. 6.18). The

---

Problem 6G   *An unusual tessellation*

Most tessellations cover the plane once. Every point of the plane is either inside one tile of the tessellation, or on an edge. What happens to the tessellation in Fig. 6.17 as more tiles are added?

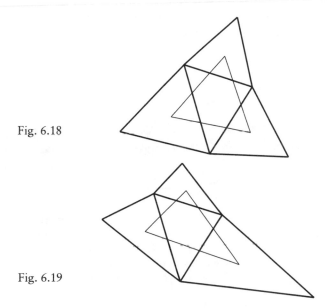

Fig. 6.18

Fig. 6.19

theorem is now *Napoleon's theorem*, named after the French emperor, who was an amateur mathematician. This says that the centres of equilateral triangles constructed on the sides of an arbitrary triangle form another equilateral triangle. (Once again, if we can see this as a small portion of a 'spiral similarity' tessellation, as in Fig. 6.17, then we can say at once that Napoleon's theorem is obviously true.) This is more than a method of proving Napoleon's theorem, without any calculation (*provided* that you know our original theorem): it also provides a generalization of Napoleon's theorem: 'Construct three similar triangles on the sides of an arbitrary triangle, oriented as in Fig. 6.16; then any three corresponding points will form a fourth similar triangle.'

The processes of specialization and generalization are opposites, as it were. From any theorem, it is possible to go to *special* cases. This is usually a relatively easy process. It is also possible to go from special cases to much more *general* theorems, though it is only an article of faith among mathematicians that such general theorems always exist.

Much of mathematics is a search for general theorems to explain special cases. The mathematician-as-scientist is frequently looking for general patterns in collections of data, which can be easy if there are lots of data, hard if there are few. Finding generalizations of a single

Problem 6H   *More equilateral triangles*

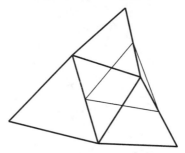

Fig. 6.20

With the benefit of hindsight, we can see that in Fig. 6.18 there are actually many more equilateral triangles, in fact an infinite number. All we have to do is to compare Fig. 6.18, which is *misleadingly* symmetrical, with the generalized theorem. What other sets of points in Fig. 6.18 form equilateral triangles? (One answer is illustrated in Fig. 6.20)

theorem may be very hard indeed. Specialization and generalization have another feature: they can often be done in several ways. Figure 6.21 illustrates another way of looking at our original theorem. The lower vertices of the pair of bold triangles have been chosen to be the start of two square grids, of different sizes. The theorem says that the free vertices will form the start of another square grid, also of a different size, and orientation.

Fig. 6.21

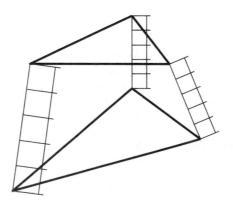

We can complete these grids (or imagine them completed, because they will go on for ever): the theorem then implies that, if we join three corresponding points, one from each grid, then they will always form triangles of the same shape. In Fig. 6.22, we have joined the corresponding points (2, 1) to form a triangle similar to the original triangles. If we pick three moving points, then corresponding positions will form triangles of the same shape as they move.

Fig. 6.22

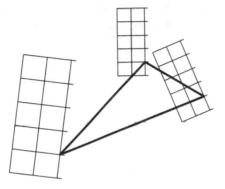

Problem 6I   *Inscribed equilateral triangles*

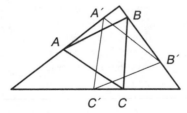

Fig. 6.23

We saw on p. 20 that, if the 'average' of two similar triangles (in the same orientation) is found by joining corresponding vertices and marking the midpoints of the resulting segments, then the 'average' triangle is also of the same shape.

Suppose that we take two equilateral triangles inscribed in a given triangle (Fig. 6.23) and imagine that vertex A is moving steadily to A' in the same time that B takes to move at constant speed to B', and C takes to move at a (different) constant speed to C'. What can be said about the intermediate positions of the three points? What can be concluded about the number of equilateral triangles that can be inscribed in a given triangle?

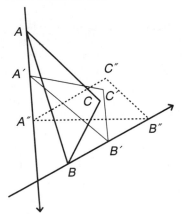

Fig. 6.24

In Fig. 6.24, we have imagined that two points move along the corresponding heavy lines; we can think of these as the y axes of two of the grids in Fig. 6.22. If we remove the grids, we get the theorem that, if two vertices of a triangle move along fixed lines each at a constant speed (different for each vertex) so that the shape of the triangle remains the same while its size changes, then the third vertex also moves at constant speed along a straight line; in Fig 6.24, C, C′ and C″ are equally spaced along a third straight line. Likewise, if two vertices of a square move at constant speeds along two lines, so that the square remains a square, the other vertices will also move at constant speed on straight lines. More generally, if any two points of a figure move along similar paths, whether curved or straight, each at a constant speed, so that the figure does not change its shape, then any other point at all in the figure will trace out a similar path, also at constant speed.

Not only theorems can be generalized. An idea can be generalized also. Figure 6.25 shows a circle and a tangent to it. A tangent might seem quite a general sort of idea already. Every smooth curve has a tangent at every point. How can this be made any more general?

Fig. 6.25

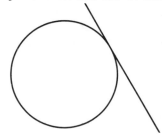

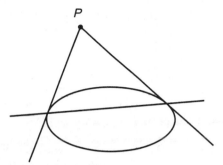

Fig. 6.26

Nineteenth-century geometers discovered how: they took a point, such as P, off a curve like the ellipse shown in Fig. 6.26 and drew two tangents to the curve. Then they joined the points of contact and called this line the *polar* of the original point, P, which is the *pole*. As the point P gets closer to the curve, the two tangents almost coincide, as in Fig. 6.27. When P is on the curve, the two tangents do coincide, into the single tangent at P, forming a special case of the pole-and-polar.

Having generalized the idea of a tangent, they were then able to discover many properties of poles and polars. The simplest is illustrated in Fig. 6.28. If P lies on the polar of Q, then Q lies on the polar of P.

Fig. 6.27

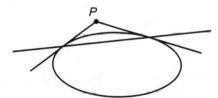

Fig. 6.28

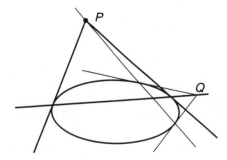

Next, they were able to do something quite remarkable (but also typical). They were able to define the polar line of a point *inside* an ellipse, despite the obvious fact that you cannot draw any real tangent from such a point to the curve. This is yet another example of generalization. The argument is both simple and elegant. If P is a point inside the ellipse (see Fig. 6.29), then the tangents at the points where *any* line through P cuts the curve will meet in a point, call it Q, such that 'P lies on the polar of Q'. It follows that Q is one point on the polar of P.

Taking a second line through P and drawing the two tangents to define a second point Q', we have *two* points on the 'polar' of P.

Fig. 6.29

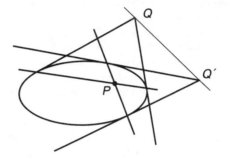

Problem 6J   *Poles and polars*

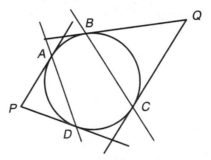

Fig. 6.30

Figure 6.30 shows a circle with two poles and polars drawn. What property do the poles P and Q and points of tangency A, B, C and D possess?

Therefore, if the polar of P exists (although P is inside the ellipse), and if the polar is a straight line, then it must be the line QQ'. In fact, experiment shows (and it can also be proved) that whatever line through P is used to construct a point Q'', Q'' will indeed lie on the line QQ', so it makes perfect sense to say that this is indeed the polar of P.

In fact, if *imaginary* points are allowed, and given coordinates by using complex numbers, then it turns out that there are two imaginary tangents from P to the curve, and QQ' is the line through their imaginary intersections with the ellipse.

The same kind of generalization occurs in arithmetic and algebra. A very important example occurred, again in the nineteenth century, as a result of attempts to solve Fermat's Last Theorem. At a meeting of the Paris Academy in 1847, Gabriel Lamé enthusiastically announced that he had proved the theorem, by a method which was related to the factorization of $x^2 + y^2$ into $(x + iy)(x - iy)$.

Unfortunately for poor Lamé, his friend Liouville rose as soon as Lamé had sat down, pointed out that Lamé's idea wasn't very new, and anyway probably did not work. Lamé had assumed, in effect, that complex numbers can be factorized just like ordinary integers. Euler had made this assumption in a few special cases, but was it a sound assumption? No, it was not.

We can illustrate the problem with an example: first, with ordinary integers. The integer 24 has several factors, and we can write it, for example, as

$$24 = 2 \times 12 = 6 \times 4 = 3 \times 8.$$

If we pick out a particular prime factor, 3 say, then we can be quite certain that 3 divides either 2 or 12, and either 6 or 4. And of course this is true. Now let's see what can happen with complex numbers. Take the number 21, which is 3×7. Calculation shows that this number can also be expressed as the product of pairs of numbers of the form $a + b\sqrt{(-5)}$:

$$21 = 3 \times 7 = [4 + \sqrt{(-5)}][4 - \sqrt{(-5)}]$$
$$= [1 + 2\sqrt{(-5)}][1 - 2\sqrt{(-5)}].$$

The last two expressions are each 'the difference of two squares', so that the first of them, for example, is equal to $4^2 - [\sqrt{(-5)}]^2 = 4^2 + 5 = 21$.

At this point, Lamé in effect assumed that $4 \pm \sqrt{(-5)}$ are 'complex integers', much like ordinary integers, and that 3 would divide into

---

Problem 6K   *An impossible factorization*

If $1 + 2\sqrt{(-5)}$ were a factor of $4 + \sqrt{(-5)}$, then there would be another complex integer, $a + b\sqrt{(-5)}$, such that

$$4 + \sqrt{(-5)} = [1 + 2\sqrt{(-5)}][a + b\sqrt{(-5)}].$$

Why is this not possible?

---

either $4 + \sqrt{(-5)}$ or $4 - \sqrt{(-5)}$, but this is not so, because both of $4 + \sqrt{(-5)}$ and $4 - \sqrt{(-5)}$ are 'primes' among numbers of the form $a + b\sqrt{(-5)}$. They cannot be written as the product of two numbers of that form. Indeed, among numbers of the form $a + b\sqrt{(-5)}$, all these factors of 21 are prime numbers! The result was that mathematicians were forced to generalize their idea of a prime number, so that eventually their more general and more powerful conception included ordinary integers, and numbers of the form $a + b\sqrt{(-N)}$.

It might seem that our perception of geometrical figures or objects will naturally be clear and easy, but this is not so. This is illustrated perfectly by the difficulty that many people have with the knight's move in chess (Fig. 6.31). The knight moves one space like a rook in any direction, and then two spaces like a rook at right angles to the original direction: or you could say instead that it moves one space like a rook and then one space like a bishop, moving away from its original position. Put in so many words, it is perhaps not surprising that it takes getting used to. Experienced players 'see' the knight's move instantly, and yet it can still be tricky to see and remember the

Fig. 6.31

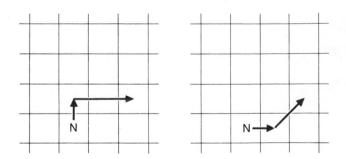

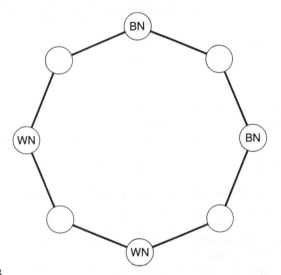

Fig. 6.32

Fig. 6.33

results of a sequence of moves. In this puzzle, posed by Guarini in 1512 and illustrated in Fig. 6.32, the object is to make the black and white knights exchange places, in as few moves as possible. Not difficult, but it becomes trivial if the eight squares which the knights can possibly occupy are strung out in a circle, as in Fig. 6.33. The original knight's moves, which require a careful act of perception to follow in sequence, are reduced to trivial steps round a circle. The solution is now so clear that it is no longer a puzzle.

Rather more complicated puzzles can be clarified in a similar manner. The reader probably remembers the Rubik's cube mania. Not long after the craze started, variants appeared, trying to cash in on its success. One variant consisted of three rings of small spheres, each of which could be rotated, which was perhaps invented by thinking of just three adjacent faces of Rubik's cube, around one

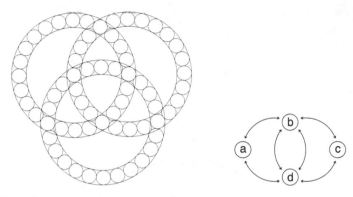

Fig. 6.34

corner. Figure 6.34 shows just three such rings on the left, and on the right, the simplest possible non-trivial version of the three rings puzzle, with only two rings and only three balls in each ring, two of which occupy the sites where the rings overlap. One move consists of rotating either of the rings through 120°, either clockwise or anticlockwise.

In order to understand this puzzle clearly, we can (because it is so very simple) actually make a list of all the possible positions of the four balls; we can then draw a graph, in which each vertex represents one position, and two vertices are joined with a line if it is possible to pass from one position to the other in one move. Figure 6.35 shows the resulting graph. There are just twelve possible positions, and they can be arranged symmetrically at the vertices of this truncated cube, each of whose triangular faces represents a sequence of three positions obtained by rotating one side of the puzzle three times in the same direction, returning to the original position. It is now very easy indeed to see how to get from any given position to any target

---

Problem 6L *A Hamiltonian puzzle*

Is it possible to find a sequence of moves on the graph in Fig. 6.35 which will start at a given position and pass through all twelve positions once, without visiting any position twice? Can it be done by a path that returns to its starting point? This is equivalent to solving Hamilton's problem for this truncated cube. (See p. 141.)

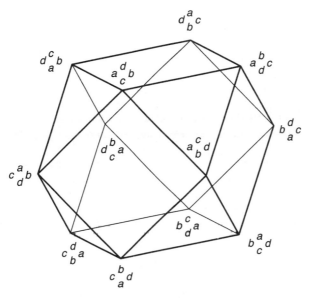

Fig. 6.35

position in the smallest number of moves. It is only somewhat harder to decide how to move between positions in larger numbers of moves, or to tackle other questions which would be, at the least, much harder with only the original puzzle in front of us.

This way of looking at the puzzle, however, puts the original Rubik's cube in perspective. There are 43,252,003,274,489,856,000 different attainable positions on the original, and while these can, in theory, be arranged to form a symmetrical graph in a similar manner, there is no prospect at all that scanning the graph with our eyes will allow us to pick out solutions to the puzzle! To picture the scale of such a graph, suppose that it could be arranged on the surface of a sphere: if the vertices of the graph, one for each position, each occupied on average an area of just one square centimetre, then the diameter of the sphere would be more than 37,000 kilometres, or about three times the diameter of the earth. Yet Rubik's cube is physically no larger than a human hand, and the puzzle can be solved from a random starting position by a skilful solver in a matter of seconds, in a few dozen moves at most.

The key to the analysis and understanding of Rubik's cube is to label certain simple moves, such as a half-turn of one face, and then to consider sequences of moves represented by their labels. Using the standard labels, the six faces are labelled *up*, *down*, *right*, *left*, *front* and *back*, and each face can be given a clockwise twist, an anticlockwise twist, or it can be rotated through 180°. (For the front face these would be denoted by F, F' and F^2.) We can now label sequences of moves such as $F'RFR$ and $RUR^2FRF^2UFU^2$.

What we have here is a kind of algebra, and the fact that in this case algebra is so powerful where geometry fails us illustrates an important feature of mathematics, and the way that our brains work. In situations which are sufficiently simple, a picture is the clearest possible way to show us what is happening, which is why mathematicians draw graphs of functions, use geometrical diagrams in two dimensions, and pictures or models of three-dimensional objects, and even use computer images to give them some idea of what objects in four and more dimensions 'look like'. What is more, such visual images can show us features which are very difficult to describe in symbols.

Yet, when the situation becomes too complicated, our sense of sight – our visual imagination – gives up, and we are forced to resort to symbols and algebra, which prove to be very effective. In fact, all the possible moves on Rubik's cube form a *group*. We can illustrate what a group is by, once again, going to a simple example. Figure 6.36 shows a shallow equilateral triangular tray holding one tile. On the tile is an arrow. On the right, the tile has been placed incorrectly in the tray: the dotted figure indicates that the arrow is now on the back of the tile. Our puzzle is to return the tile to its correct position using two types of move. We can rotate the tile clockwise through an angle of 120°, move A, or through an angle of 240°, denoted by AA or A^2; or we can flip it about one of the lines of symmetry of the triangle. These have been labelled P, Q and R in Fig. 6.37.

Fig. 6.36

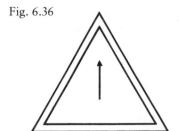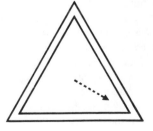

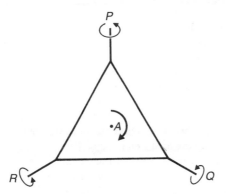

Fig. 6.37

In the second illustration, we can solve our problem by giving the tile a rotation *AA*, and then a twist, *P*; or we could give it a twist *P* at once and then a rotation of only *A*; or we could give it a twist *R* and nothing more; or we could give it a twist *Q* and then a rotation of *AA*, . . . Clearly, there are many ways to solve the puzzle, or to get from one position of the tile to another. To clarify the situation, the table in Fig. 6.38 shows the results of every possible pair of moves. Notice that we have introduced an *identity* move, *I*, which does nothing, but leaves the tile as it is. This is a *group* table, and it defines the particular group based on the moves *I, A, A², P, Q* and *R*.

The first operation to be performed is shown along the *top* of the table, and the second down the side. It is a convention that, when operations are written in sequence, the first is placed to the right, so that *PA* means that *A* is done first, and then *P*. The notation *PAQ* means *Q*, then *A*, then *P*.

Fig. 6.38

First Operation

	I	A	A²	P	Q	R
I	I	A	A²	P	Q	R
A	A	A²	I	R	P	Q
A²	A²	I	A	Q	R	P
P	P	Q	R	I	A	A²
Q	Q	R	P	A²	I	A
R	R	P	Q	A	A²	I

Second Operation

Checking the table, or experimenting with an actual tile, will show that the table has four properties: every pair of operations is equivalent to another (single) operation; there is an operation, I, which leaves the triangle unchanged; for every operation, there is an *inverse* operation which reverses its effect; and for any three operations – let's take P, A and Q as an example – we have $P(AQ) = (PA)Q$. In other words, if we perform AQ and then P, we get the same result as if we perform Q then PA. These are the properties that mathematicians use to *define* the idea of a group.

The group table represents all the essential features of the original puzzle. Moreover – and this is why it is of such interest to the mathematician – it also represents the essential features of many other situations. Comparing the table with the original puzzle, we can see that there are six possible operations and six possible positions of the tile. Since every operation leaves the tile in a different position, it follows that we can get to any other position we choose in just one move – rather a contrast with Rubik's cube.

That was a fact about the puzzle which we deduced from the table. However, the table shows us much more general facts. Even if we had never seen the original puzzle, we could see striking patterns in the table itself. Firstly, it divides obviously into quarters, and the top-left quarter refers to nothing but I, A and A^2. This mini-table represents the effects of rotation through 120°, and would apply just as well to a wheel as to an equilateral triangle. We can also 'see' it as representing another situation. Figure 6.39 is an addition table for the integers 0, 1 and 2, with the condition that when the total goes over 2, we take the remainder on dividing by 3; so $2 + 2 = 4 = 3 + 1 = 1$. On reflection it is no surprise that this kind of addition has the same pattern as the rotation of a wheel, because we can represent it as adding, or counting, round a wheel, arranging the numbers 0, 1 and 2 as in Fig. 6.40. In fact, the three elements I, A and A^2 form by themselves a self-contained group, which is described as a *subgroup* of the original group.

Fig. 6.39

	0	1	2
0	0	1	2
1	1	2	0
2	2	0	1

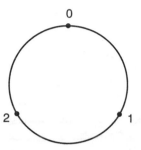

Fig. 6.40

Relations such as these hint at the profound importance of group theory in mathematics. Again and again mathematicians have found the same group patterns – in the solution of equations, in geometrical transformations, in the study of permutations (which happens to include campanology or bell-ringing), in the topological study of surfaces in which each surface is associated with a group, and in many other areas. In chemistry, groups are used at an elementary level to study the possible forms of crystals; at a higher level, group theory is used to represent the symmetries of complex molecules and to simplify the complicated equations that describe them, so that they can be solved. In physics, conservation laws can be expressed in terms of symmetries and therefore are also related to groups.

In all these cases, the group pattern is a way of expressing the fact that apparently different situations are actually closely *analogous*. The same (group) pattern underlies them: we turn to this important kind of perception in the next chapter.

Solutions to problems, Chapter 6

6A: *Traditional riddles*

A coffin; fire; your breath; your name; silence; there is not a single person in it; the first is going to itch, the second is itching to go; moths, which eat holes; giant footprints all over Australia.

6B: *Mathematical riddles*

There isn't a largest fraction or irrational number less than 1.

The simplest solution is plausibly 'equivalent fractions', which cause so much trouble to young pupils, because of these very properties.

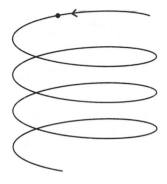

Fig. 6.41

We have already considered the 'centre of a triangle'.
One.
Two or zero. Early Greek mathematicians, for this and other reasons, were uncertain as to whether 2 was a number at all. ('One' certainly was not – it was the original unity and the generator of all other numbers, but not a number itself.) Zero was invented centuries later in India.

The simplest answer is plausibly the cuboctahedron, which is a truncation of either the cube or the octahedron by slicing them through the midpoints of adjacent edges surrounding each vertex. (It is the polyhedron used on page 191 to map the graph of the twelve possible positions in the ring puzzle.)

A point on a circle returns to its starting point after one revolution, but a point on a helix never returns. The helix (Fig. 6.41) is the path of a point which rotates steadily around a cylinder while also moving steadily along it.

6C: *Slicing a cube*

Figure 6.42 (left) shows the regular hexagonal cross-section. It slices through the midpoints of six of the twelve edges of the cube. If eight cubes are stacked to make a larger cube, and each cube is sliced to show a regular hexagonal face, all these faces facing outwards, then the result is eight pieces forming a truncated octahedron. In contrast to the cuboctahedron of Problem 6B, the octahedron has only been sliced at its vertices far enough to make every vertex into a square and every face into a regular hexagon, as shown in the right-hand

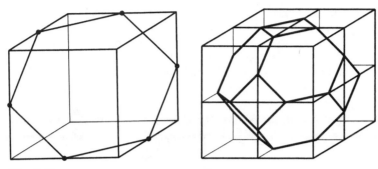

Fig. 6.42

diagram of Fig. 6.42. Since the cubes originally filled space, and every half-cube is now one eighth of a truncated octahedron, the new solids fill space also.

6D: *The symmetry of the cubic*

By observation, $2 \times 3 \times 4 = 24$ and $2 + 3 + 4 = 9$. Combining the roots two at a time gives $2 \times 3 + 3 \times 4 + 4 \times 2 = 26$. In general, if the cubic has roots a, b and c, then it can be written in the form

$$x^3 - (a + b + c)x^2 + (ab + bc + ca)x - abc = 0.$$

Similarly, if the roots of a fourth-degree equation are a, b, c and d, then the coefficients, ignoring their signs, are: $a + b + c + d$, $ab + bc + cd + da + ac + bd$, $bcd + cda + dab + abc$ and $abcd$.

6E: *The smallest triangle*

Consider the fence as a movable line, which is rotating about the point W (Fig. 6.43). Then W divides the line into two parts. If one of

Fig. 6.43

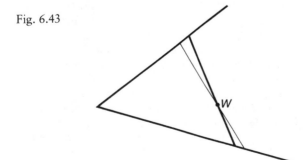

those parts is longer than the other, then rotating the line slightly will make them more equal in length, *and decrease* the area, because the area lost from the longer part of the line will be greater than area gained by the shorter part. Therefore, the area of the triangle will be a minimum when W is the midpoint of the fence.

6F: *Cardano's spur-wheel problem*

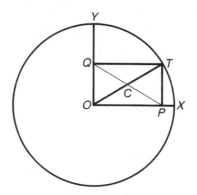

Fig. 6.44

Consider a radius, OT, of the large circle rotating through one quarter of a complete circle, starting at OX and ending at OY (Fig. 6.44). It is a diagonal of a rectangle $OPTQ$ whose other diagonal is PQ. As OT rotates, P and Q slide along the radii OX and OY.

Consider the circle on PQ as diameter, whose centre is C, which will always be touching the larger circle. Initially, when QP coincides with OX, it will touch the larger circle at P, and finally, when QP coincides with YO, it will touch the larger circle at Q, the other end of the same diameter. It follows that one half of this circle's circumference has been in touch with the larger circle. However, the smaller circle, being one half of the diameter, also has one half of the circumference of the larger circle, and so one half of its diameter is equal to one quarter of the diameter of the larger circle. Therefore the circle on PQ as diameter is rolling without slipping on the larger circle.

It follows conversely that, if the smaller circle rolls without slipping on the larger, then the end of a diameter traces a radius of the larger circle as the smaller circle completes one quarter. During the complete rotation of the smaller circle, the end of a diameter traces a diameter of the larger circle in both directions.

Note: Cardano's original problem, which is discussed in H. Dorrie's *100 Great Problems of Elementary Mathematics: Their History and Solution*, asks for the path of a point anywhere on the rotating wheel. Since any point on the wheel lies on a diameter, and the movement of the diameter is that of a slipping ladder, the movement of a general point is the path of a point on such a ladder, which is known to be an ellipse.

6G: *An unusual tessellation*

The tessellation starts to overlap itself; then, as more tiles are added, it overlaps itself again, and again, and again. In fact it overlaps itself an infinite number of times.

6H: *More equilateral triangles*

The figure for Napoleon's theorem is given in Fig. 6.45, with arrows to show the orientation of the triangles, as required by the generalization. Any triplet of three corresponding points will now form an equilateral triangle; we have marked several simple sets.

Fig. 6.45

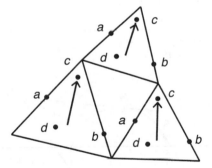

6I: *Inscribed equilateral triangles*

Any intermediate position of the three moving points will also form an equilateral triangle (Fig. 6.46). Consequently, there are an infinite

Fig. 6.46

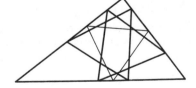

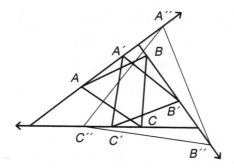

Fig. 6.47

number of equilateral triangles inscribed in a given triangle, provided there are at least two. (This argument does not however establish that there are two, or even one.) If we follow the same three moving points as they move along the sides of the triangle extended beyond the vertices (Fig. 6.47), they will still form the vertices of an equilateral triangle, so we might consider that these were 'inscribed' also, since their vertices lie on the sides of the triangle, although they generally lie outside the triangle, not inside it.

6J: *Poles and polars*

The lines *PQ*, *AC* and *BD* concur, as illustrated in Fig. 6.48.

Fig. 6.48

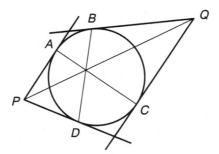

6K: *An impossible factorization*

If it were possible, then the rational parts of each expression would be equal, and also the irrational parts. In other words, we would have

$$4 = 1a - 2 \times 5b \quad \text{and} \quad \sqrt{(-5)} = b\sqrt{(-5)} + 2a\sqrt{(-5)}.$$

That is,

$$4 = a - 10b \quad \text{and} \quad 1 = b + 2a,$$

from which we get $b = -\frac{1}{3}$ and $a = \frac{2}{3}$; so a and b are not integers, and $a + b\sqrt{(-5)}$ is not a complex integer.

6L: *A Hamiltonian puzzle*

Yes, it is. One such path is traced on Fig. 6.49 by the numbers 1 to 12. Note that, in this path, two rotations of the right-hand circle are followed by two left-hand rotations, and so on.

Fig. 6.49

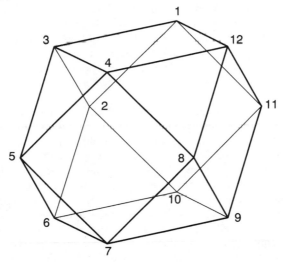

Chapter 7
Likeness and analogy

'Discovery always proceeds by similitude'
Francis Bacon

Bacon exaggerated slightly, perhaps, but excusably. Analogy lies at the heart of mathematical creativity. There was a striking analogy between Pascal's triangle and the simplest one-dimensional cellular automaton, as well as between the possible positions in the simplest variant of Rubik's cube and the vertices of a polyhedron. We have seen an analogy between several popular puzzles which are all equivalent, when suitably viewed, to puzzles about graphs, and between rates of change and the size of errors when an observation is not quite accurate.

We have seen a simple but extensive analogy between averages and centres of gravity, and points added like numbers, and weighted averages treated as points on a line. It should be no surprise that this analogy can be extended, for example to the mixing of hot and cold liquids. On the left of Fig. 7.1, equal quantities of a liquid, at two different temperatures, are mixed. The temperatures are initially 100°C and 60°C, and the mixture is naturally at 80°C, represented by the midpoint. However, on the right of the figure, a quantity of the 60°C liquid is mixed with twice the volume of the 100°C liquid, and the final temperature is now the weighted average, as marked.

The simplest analogies occur when we realize that two problems that appear at first sight to be different are in fact essentially the same. Johannes Müller, named Regiomontanus after the Latin translation of Königsberg, his city of birth, later made famous by Euler,

Fig. 7.1

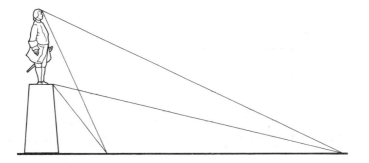

Fig. 7.2

proposed this problem in 1471. 'At what point on the earth's surface does a perpendicularly suspended rod appear largest?' The point of this problem may not appear immediately obvious, so it is usually put in this form – itself an example of using analogy to make a problem clearer or more vivid: 'From what distance will a statue on a plinth appear largest to the eye?' (see Fig. 7.2). If we approach too close, the statue appears foreshortened, but from a distance it is simply small. It is not difficult to see that this is similar to the problem of where to sit in the theatre to get the widest view of the stage. A box in the wings may be prestigious, but you see the stage from a narrow angle (Fig. 7.3), while the seats in the gallery are just a long way away.

The next – modern – version of this problem is less obviously similar. A try has just been scored in the game of rugby football. The rules state that the conversion must be attempted from a point on the line at right-angles to the base line, in line with the point where the

Fig. 7.3

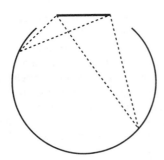

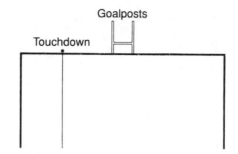

Fig. 7.4

try touched down (Fig. 7.4). If the touchdown is not between the posts, where should the kicker place the ball?

*

Problem 7A   *Searching for the widest angle*

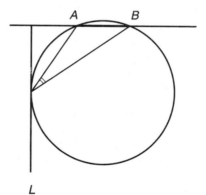

Fig. 7.5

The solutions to Regiomontanus' problem and its two analogues are identical, as we would expect. In Fig. 7.5 we are looking sideways at the statue, or looking down on the theatre or the rugby pitch. The span from *A* to *B* represents the proscenium arch, or the statue or the goalmouth. Line *L* is either the ground on which the statue's plinth rests, or the line of the rugby try. It does not represent an exact line in the theatre analogy, only a rough guide. To maximize the angle subtended by *AB*, we draw the circle through *A* and *B* that just touches *L*, and we stand just where it touches. Why does this solution guarantee us the best view?

A well-known puzzle concerns a tennis tournament. It is arranged with 32 players, so that there will be 16 in the second round, 8 in the third and so on, but unfortunately three extra players turn up, who must be accommodated. The draw will now contain byes (which need not be necessarily in the first round). How should the byes be arranged to minimize the number of matches played?

Compare this puzzle with the following two: in the first, you are assembling a jigsaw puzzle with 500 pieces. As you search for the piece with a bit of sky and a bit of red rose, you wonder whether it is a better strategy to build up from one piece, making it larger all the time, or whether you should not start building several small islands, as it were, and then have the satisfaction of joining whole islands together sooner or later. To be specific, if joining two portions of the jigsaw together counts as one move, whether a portion is a single piece or a number of pieces joined together, how should you play to minimize the number of moves required?

While doing the jigsaw, you are eating a large bar of chocolate, divided as usual into small squares. In fact, it is divided into six rows by four rows of identical squares. You eat one square at a time, and curiosity prompts you to wonder how few breaks are needed to divide the whole bar into individual squares, supposing that you do not try to be clever and break several separate pieces in your hands at one go. All three puzzles have the charm of being rather natural: they are also perfect analogues of each other.

Abandoning the delights of tennis, imagine that you are a forest ranger and that you wish to estimate the number of a certain small animal on the range. To catch them all is out of the question, but it is reasonable to catch a sample, and mark them. You can release them, allow them time to mix thoroughly with the rest of the population, and then capture a second sample, which you examine to see how many were marked from the first time round. Suppose that the total population is P, that you first capture and mark F animals and then capture S animals of which B have already been marked. If the animals were really thoroughly mixed, it is a reasonable assumption that the proportion of marked animals in your second sample is equal

---

Problem 7B   *How many matches, moves or breaks?*

What are the identical solutions to the last three analogous problems?

to the proportion of all marked animals in the total population. So, $F/P = B/S$ and P is estimated at FS/B. This is known as the *Lincoln index*.

You happen to have a friend who is a printer, who employs two proofreaders, because no proofreader spots every error. If the proofreaders check the same proof, and the first finds 30 errors, and the second 24 errors, of which 20 errors are common to both, how many errors does the whole text plausibly contain? If the total number of errors is E, then the first reader spots $30/E$ of them, and the second spots a smaller proportion, $24/E$. Assuming that the second spots the same proportion of errors, over the whole text, as he spots among the errors spotted by the first reader, then $20 = 24/E \times 30$, from which E is estimated as $30 \times 24/20$ or 36 expected errors.

These problems are, in effect, identical, or as mathematicians might say, *isomorphic*, from the Greek for 'having the same form'. The animals scattered through the forest correspond to the errors scattered through the text. In each case, two sets of animals/errors are picked out and counted, and then compared for the number of animals/errors in common. The same conclusions follows in each case.

Dice pose many puzzling problems. To solve them, it often helps to realize that it makes no difference whether several dice are thrown in sequence, or all thrown together, or whether one die is thrown again and again, provided that there is no connection between the different throws: they must all be independent and not affect one another.

Now consider the probability that two integers picked at random from, say, the first billion integers, have no prime factor in common. It turns out that the probability is approximately $6/\pi^2$. (The number π does turn up in some unexpected places.) Now consider the probability that a number, chosen at random in the same way, does not contain the same prime factor twice (or more). Such numbers are called *square-free*. The number 12 is not square-free because it is divisible by 2^2, while 30 is square-free, because its prime factors, 2, 3 and 5, occur only once each.

The probability that a number is square-free is, as it were, the probability that it does not share a prime factor with itself. If we imagine that we are throwing a large die, on whose faces are all the prime numbers which could divide a number less than a billion, then it makes no difference whether we are throwing it twice to decide on the prime factors of one chosen number, or whether we throw it twice, once each for two different numbers. When mathematicians make this analogy with dice throwing more precise, it does indeed turn out that the probability that a number is square-free is $6/\pi^2$.

Going on holiday to the seaside, you spend the afternoon on the beach, which consists of an expanse of sharp pebbles, followed by firm sand, and then the water. Having enjoyed your swim, you wish to return to your deckchair in minimum time (see Fig. 7.6). You consider crossing the sand to X and taking the shortest route across the pebbles. However, by heading for somewhere nearer to Y, you would walk markedly less on the sand and only marginally further across the pebbles. Where is your minimum route?

This problem is roughly solved by eye, every year, by thousands of holiday makers. It appears in later editions of Jacques Ozanam's eighteenth-century *Recreations in Mathematics and Natural Philosophy*, where the puzzle concerns a horse that is trotting first over soft sand, then over fine turf.

If you remember some elementary physics from school, you may realize that the problem has already been solved, provided only that you want to get to your deckchair as quickly as possible and that you know your relative speeds across the pebbles and sand. Replacing the rough boundary between the sand and the pebbles with the smooth surface of a volume of water, it becomes a problem in refraction (Fig. 7.7): the speed of light is less through water than in air, and so the ray of light, which always takes the fastest route between two points, is bent as it enters the water. According to Snell's Law, $(\sin a)/\sin b$ is constant, the constant depending on the two media, air and water.

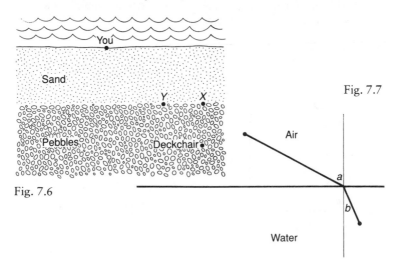

Fig. 7.7

Fig. 7.6

Problem 7C   *Bouncing on water*

Is there an analogy between rays of light which strike the surface of water at a shallow angle, and objects which do the same?

---

*

The same edition of Ozanam's work presents as Puzzle XXXIII: 'A gentleman wishes to have a silver vessel of a cylindric form, open to the top, capable of containing a cubic foot of liquor; but being desirous to save the material as much as possible, requests to know the proper dimensions of the vessel.'

A possible analogue comes from a very different source: *The Ladies' Diary or Woman's Almanac* was first published in 1704. It initially published articles on cooking, health and education, but soon changed to a journal of enigmas, word puzzles and mathematical questions. Women continued to pose and solve many of the questions, though naturally men could not keep their fingers out of this attractive pie. This question was posed by Mrs Barbara Sidway: 'From a given cone to cut the greatest cylinder possible.' Does it in any way resemble Ozanama's open-cylinder problem? (see Fig. 7.8). The total surface area of Ozanam's vessel is $S = \pi r^2 + 2\pi rh$, where r is the radius, and h is the height, of the cylinder. Its volume is $\pi r^2 h$. To minimize the amount of silver used, given a fixed volume, is essentially the same as maximizing the volume, for a given quantity of silver. So this is one resemblance to the cylinder-in-the-cone problem; they can both be seen as *maximum* problems.

Fig. 7.8

Ozanam's silver vessel   Mrs Sidway's cone

If the area of Ozanam's vessel is fixed, then $S/\pi = r^2 + 2rh$ is fixed, and, given that condition, we have to maximize $V/\pi = r^2h$. How do we do this? By recalling the idea that if the sum of several variable quantities is fixed, then their product is a maximum when (and if) they are all equal.

With this idea in mind, we try to rearrange the two terms in the expression for S so that their product somehow matches r^2h. Can we? Yes, by writing

$$S = r^2 + rh + rh$$

and noticing that $r^2 \times rh \times rh = r^4h^2 = (V/\pi)^2$. Therefore, $(V/\pi)^2$, and therefore V/π, and therefore V, will be a maximum when $r^2 = rh = rh$ which will happen when $r = h$. This is the solution; the height of the cylinder must equal its radius.

Can the *Ladies' Diary* problem be solved in the same way? The volume of the cylinder is proportional to $DE^2 \times PQ$, and we know that DE/AP is constant. Therefore, it is required to maximize $AP^2 \times PQ$. In other words, given any line AQ, find a point on it, P, such that $AP^2 \times PQ$ is a maximum. Since $PQ = AQ - AP$, we have to maximize $AP^2(AQ - AP)$, which is the same as maximizing $AP^2(2AQ - 2AP)$ which can be thought of as the product of three factors: $AP \times AP \times (2AQ - 2AP)$.

Fig. 7.9

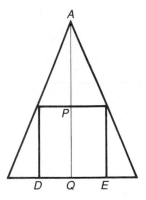

This product is similar to that in the Ozanam problem. The sum of the three terms is fixed, at $2AQ$. The product will be a maximum when $AP = 2AQ - 2AP$, or $AP = \frac{2}{3}AQ$.

*

Problem 7D   *Rabbi Ben Moses' square vessel*

Jewish dietary laws traditionally divide food into the pure and the impure. However, the addition of a very small proportion of impure food does not make a pure food impure. It follows that a cooking vessel whose walls might have in the past absorbed some impurity can be used safely, provided the ratio of its volume to the total area of its walls is sufficiently large. Consequently, many problems of relative area and volume were analysed by Jewish philosophers and theologians. In particular, Rabbi Solomon Ben Moses solved in the eighteenth century the problem of finding the box, with a square base and open top, whose volume in proportion to the surface area of the sides and base is a maximum.

How closely analogous is this problem to Ozanam's?

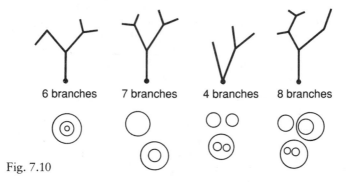

Fig. 7.10

At the top of Fig. 7.10 are shown some examples of *rooted trees* with various numbers of branches, and underneath are shown some ways of arranging bubbles, either next to each other, or inside each other, or a combination of both. What is the connection between the bubbles and the trees? That there is a connection is suggested by an experimental test: carefully list all the separate trees with $n = 1, 2, 3,$ ... branches, and similarly the arrangements of bubbles containing $n = 1, 2, 3,$ ... bubbles; then the number of types of tree, $t(n)$, and the number of arrangements of bubbles, $b(n)$, are identical for small n:

$$n: \quad 1 \quad 2 \quad 3 \quad 4 \quad 5 \quad 6 \quad 7 \quad \cdots$$

$$t(n), b(n): \quad 1 \quad 2 \quad 4 \quad 9 \quad 20 \quad 48 \quad 115 \quad \cdots$$

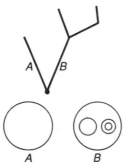

Fig. 7.11

How suspicious should we be of this match? Is this another misleading analogy, or is it genuine? We can see that it is genuine by a direct comparison. An arrangement of bubbles and the matching rooted tree are shown in Fig. 7.11. To match them, feature for feature, we have labelled both figures. We notice that there are two outer bubbles, A and B, which match the branches A and B from the root of the tree. Bubble branch A now takes no further part. Bubble (branch) B however contains (is attached to) two more bubbles (branches) one of which contains yet another bubble (is attached to another branch).

By matching the problems perfectly, we can now answer questions about one problem, by studying the other. As it happens, Cayley solved the problem of counting rooted trees in 1857, so that bubble problem is also solved!

It might seem that just *counting* a number of objects was very easy. After all, don't small children manage to count? Yes, they do, but mathematicians face difficulties that children seldom meet. Their difficulties start with the problem of listing the objects you are counting and arranging them in some sort of order, and then making sure that, when you count the different objects, every one is counted once and only once. Readers who try to verify for themselves that there are 48 arrangements of six bubbles – let alone 115 arrangements of seven – will appreciate this point from their own experience.

The difficulty of counting has given rise to a whole branch of mathematics, called combinatorics, sometimes described as the art of *advanced counting*. Fortunately, an amazing number of problems in combinatorics are exact analogues of each other. Solve one, and you solve both, or all of them.

Figure 7.12 represents a *random walk*. The horizontal axis represents the time that has passed, in whole units, with the start or origin

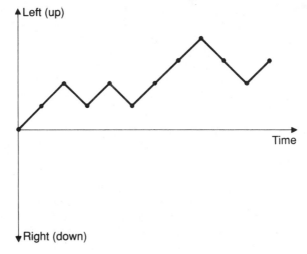

Fig. 7.12

on the left. At each step the point represented by the zigzag line moves up one unit or down one unit. There are no other possibilities. This is called a random walk because you can imagine that it represents a drunk on a straight line, staggering one step to the left, or one step to the right, but doing so randomly, possibly observed by an interested police officer.

Problem 7E *The ballot problem*

This problem was solved by W. A. Whitworth in 1878: 'In a ballot, the winning candidate P scores p votes, and the losing candidate Q scores q votes. What is the probability that during the count, P was always ahead of Q?' The answer turns out to be $(p - q)/(p + q)$. Our query is this: why is this equivalent to a random-walk problem?

Here is a problem about brackets. Imagine a long complicated algebraic expression from which everything has been removed except the brackets: for example,

$$(\ (\ ) \ (\ (\ ) \ ) \ (\ ) \ ) \ (\ (\ ) \ ) \ (\ ).$$

Simpler examples of *bracket sequences* are (( )) and (( ) (( ))). What is the connection between bracket sequences and the ballot problem?

Random-walk problems are important because they represent many other problems. For example, in statistics, it is natural to ask how many heads and tails you can expect to get if you toss an unbiased coin a large number of times, and to wonder how the heads and tails will be distributed. Is it likely that most of the heads will occur early on in the trials? If the number of heads and tails is roughly equal after many hundreds of tosses, is it likely that the cumulative number of heads will have been ahead of the cumulative number of tails for most of the 'race'?

These questions can all be interpreted as random-walk questions, by thinking of throwing a head as moving the point up one, and a tail as moving it down one. In two dimensions we can imagine a drunk staggering east and west and also north and south. This is an excellent model for the *diffusion* of a gas, so chemists want to know how far the drunk gets from the origin, and how quickly.

Figure 7.13 illustrates an important problem. We start at the point A, and consider how many random walks there are to B that do not touch or cross the horizontal axis. As so often, it is simpler to consider how many paths *do* cross the axis, and then subtract that number from the total of all paths. So: what is the number of paths from A to B that cross or touch the axis? The *reflection principle* says that this number is identical to the number of paths from A' to B, where A' is the reflection of A in the axis. This is so because there is

Fig. 7.13

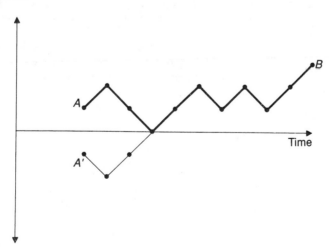

Problem 7F   *Counting random walks*

The most basic problem in random walks is simply to count the number of possible random walks between two points, A and B, or (to make it even simpler) from the origin to a given point P whose coordinates are (x, y).

Why is this problem exactly analogous to the problem of counting the number of ways of selecting p positions from a total of $p + q$ positions arranged in a straight line?

---

an exact match between paths from A (touching or crossing the axis) and paths from A'. All we have to do is to take any axis-hitting path from A, and reflect the path *up to the point where it first touches or crosses the axis*, in the axis, to get a unique path from A'. Similarly, any such path from A' can be reflected, up to when it first meets the axis, to get a unique path from A.

The effect of the reflection principle is to make many problems easier to solve. Random walks with barriers, such as axes which must not be crossed, occur frequently in practice, and are difficult to handle. By the reflection principle, a problem about just crossing the horizontal axis has been reduced to a problem with no barrier at all.

Mathematicians continually think in physical terms, of movement and change and transformation, though they know that such ways of thinking can be misleading. They sometimes rely on actual physical models, or today on images generated by a computer, or they rely on mental imagery.

Figure 7.14 shows a small heavy ring sliding on a smooth string attached to two points. The problem is to discover where the ring will come to rest. One argument is very simple: the tension in the string will be the same on both sides of the ring, and so the angles of inclination of the string on either side of the ring will be equal (see Fig. 7.15). The ring at P comes to rest where AP and BP make the same angle with the horizontal.

Fig. 7.14   Fig. 7.15

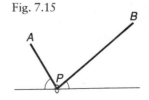

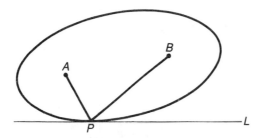

Fig. 7.16

However, there is another way of looking at this situation. The length of the string $AP + PB$ is constant, which suggests the best-known method of drawing an ellipse (Fig. 7.16): fix a piece of string with two pins and use a moving pencil to keep the string taut. So the path of P will be an ellipse, with A and B as the 'pins', technically called the foci of the ellipse. The problem has now turned into: where on the ellipse will P come to rest? Common sense says, 'At the lowest point!' This is where the ellipse just touches the horizontal line L.

Now we have two solutions, which must be equivalent to each other, so we can literally 'fit them together' and conclude that (Fig. 7.17) if an ellipse touches a line L, then the angles between the lines joining the point of contact to the foci are equal. By a small thought-experiment about a physical ring sliding on a piece of string, we have confidently inferred an abstract fact about an ellipse.

Since an ellipse can be defined as the path of a point whose sum of distances from two fixed points is constant, it is natural to wonder what the path would be if it were the *difference* between the distances which did not change. The answer is that the path is a hyperbola

Fig. 7.17

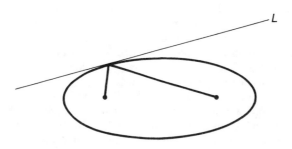

Problem 7G   *Constructing a tangent to an ellipse*

•*P*

A•

Fig. 7.18   •*B*

Using the fact just discovered, we can construct a tangent to an ellipse using just ruler and compasses, even if all we know is that the foci of the ellipse are *A* and *B*, and that *P* is a point on it (Fig. 7.18). How?

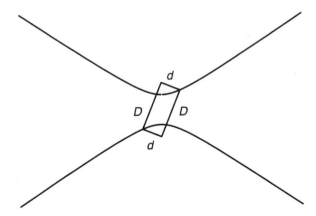

Fig. 7.19

(Fig. 7.19). Notice that, since the difference between two distances can be taken in two ways, either $D - d$ or $d - D$, as it were, we get two branches to the hyperbola, whereas the ellipse is one piece, because two numbers can only be added in one way. We immediately have an analogy: the distinct properties of addition and subtraction correspond to the two forms of the paths!

We should not be surprised that the ellipse and the hyperbola occur as answers to analogous problems, because they were first defined by Greek mathematicians as slices of a circular cone, or rather, a double circular cone (see Fig. 5.4 on p. 118). A slice that misses one half of the double cone gives an ellipse; a slice that cuts both is a hyperbola. In between, however, is a slice parallel to a line in the cone which passes through the vertex. Such a slice opens out for ever, but only in one part. It is a parabola.

Since our analogies have worked so well, so far, we wonder whether the parabola has a similar definition in terms of distances. It does (see Fig. 7.20): a parabola has one focus, and a line called the directrix, such that the distances of any point on the parabola from the focus and from the directrix are equal. In every one of these cases, the parabola comes between the ellipse and the hyperbola. It was the brilliant Johannes Kepler (1571–1630), author of the famous three laws of planetary motion, who first proposed that the parabola could be thought of as an ellipse one of whose foci had disappeared to infinity, and that the hyperbola could be thought of as the result of the same focus coming back in the other direction. This fits perfectly the idea of slicing a cone because, as the elliptical slice gets longer and longer, the plane of the slice gets closer and closer to the plane of a parabolic slice.

However, it does not fit the existence of a *directrix* for the parabola. It is not very helpful to talk of one focus of the parabola being at infinity, so the sum of the distances of any point on the parabola from its two foci is infinite, so it seems rather fortunate that the parabola has the extra focus–directrix property. Or is it? Might not the ellipse and the hyperbola also have directrices? Indeed, they do, so that analogy is complete after all. The ratio of the distances of a point on the curve from a focus and the matching directrix is

Fig. 7.20

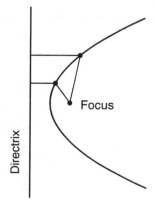

Problem 7H   *Tangents, foci and the directrix*

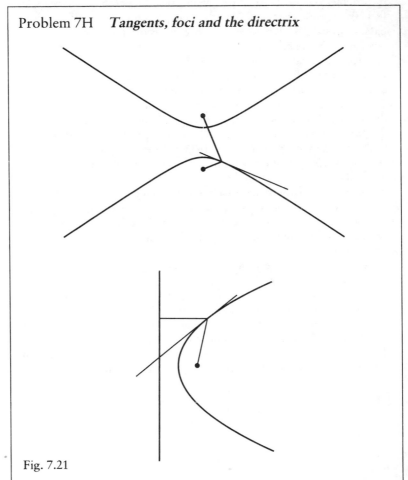

Fig. 7.21

Success breeds success. We know that the lines from the foci to a point P on an ellipse are at equal angles to the tangent at P. Are there analogous properties for the hyperbola and parabola? (See Fig. 7.21.)

constant. For a parabola, this constant is 1, for an ellipse it is less than 1 (see Fig. 7.22) and, for a hyperbola, greater than 1. As usual, the parabola marks the boundary between ellipses and hyperbolas.

*

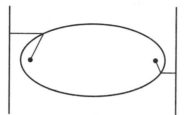

Fig. 7.22

These analogies between ellipses, parabolas and hyperbolas have all been very specific. However, the very idea of two sets of objects, ellipses and hyperbolas, separated by a set, the parabolas, which has something in common with both, is a good analogy for many phenomena in mathematics.

Figure 7.23 provides us with a picture of possible solutions to a differential equation, with the difference that whereas the solutions may be hard to calculate, such pictures are relatively easy to draw on a computer-driven plotter.

Fig. 7.23

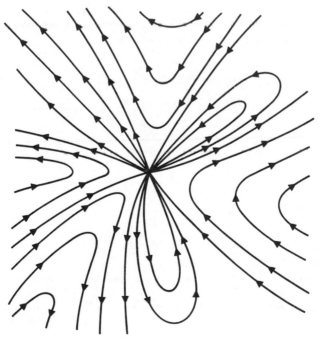

Problem 7I   *Ellipses and hyperbolas with the same foci*

This problem requires either the use of string and pins, and ingenuity, or algebra, or a computer. What is the connection between ellipses and hyperbolas with the same two points as foci?

This map contains a kind of critical point, where the lines cross, called a *focus*. The whole map can be divided into four hyperbolic sectors in which the paths approach the critical point but then veer away; two elliptical sectors in which paths start from the critical point, move away and then return; and (no surprise) parabolic paths which start at the critical point and end at infinity, and vice versa.

The importance of such maps is twofold. On the one hand, we can plot exact maps for specific differential equations, and then study them to discover features of the original equation. On the other hand, we can study the maps for their own sake, to discover what kinds of map are *possible*, to study their general features. Our results may be entertaining in themselves, and they will also tell us something about the possible solutions to differential equations – or they may tell us about quite different but analogous problems.

Problem 7J   *Plot the flow*

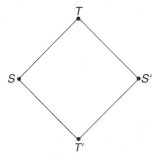

Fig. 7.24

How would you expect the flow lines to move, given the four critical points, shown in Fig. 7.24, two of which, T and T', are *sources* (explained in the text below) and two of which, S and S', are *sinks*?

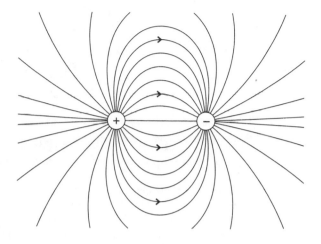

Fig. 7.25

The next map (Fig. 7.25) is relatively simpler, except that it contains two critical points, one of them a *source*, marked +, from which flow lines only emerge, and the other a *sink*, marked −, into which flow-lines disappear. This might seem to be a good picture of water flowing out of a tap and down a plug-hole, and indeed it could be, but it is actually a picture of the electric lines of force between the two ends of a magnet, a subject studied by James Clerk Maxwell (1831–79), the great scientist who unified the sciences of optics and electrodynamics by showing that the same differential equations described both. The direction of magnetic lines of force (or other fields of force due to electric currents, or gravity) can be plotted from the differential equations that describe them. From the same differential equations, we can calculate the *potential* function, and plot the lines of *equal potential*. It turns out that the equipotential lines (on each of which the potential is constant) are always at right angles to the lines of force, which brings us back to the sets of ellipses and hyperbolas with the same foci (as in Problem 7I) which are also mutually perpendicular.

These maps were drawn to represent differential equations, but there is nothing to stop the mathematician drawing a map and then looking for an equation to fit it. To illustrate why this might be useful, suppose that a scientist is studying the flow of water past a

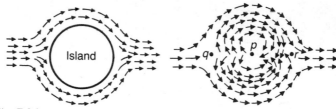

Fig. 7.26

circular obstacle, as on the left of Fig. 7.26, which roughly shows the observed flow. How could this be represented by selecting suitable critical points?

The answer is illustrated in the right-hand figure. We suppose that the flow over the whole plane is from left to right, and we place in this flow two critical points, one of which is a *source* and the other a *sink*, as if liquid were pouring out of the source and disappearing down the sink. By placing them close together, we get this pattern of flow-lines. If we now work backwards to find a differential equation which matches these flow lines, then – ignoring the central portion of the map which is replaced by the original island – we have a mathematical equation for the flow of water past the island.

We are not yet finished with our analogies! There is another analogy with – amazing to relate – Euler's relationship for polyhedra: $V + F = E + 2$. To see how it arises, imagine the surface of a minor planet divided into faces, vertices and edges, like a polyhedron. Next, imagine that water is flowing over the surface of this minor planet, and that it gushes out of a number of sources, one at each vertex, and

Fig. 7.27

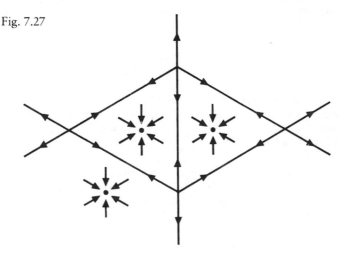

disappears down a number of sinks, one in the middle of each face (Fig. 7.27). As water flows along an edge, it will approach the middle of each edge from the vertices at either end, but then veer off to disappear down the sinks on either side. These points at the middle of each edge are called *saddle points*. Therefore, if we count V sources, F sinks and E saddle-points, then $V - E + F$ is constant. (Since most maps are actually more complicated than that simple polyhedron, the same simple result needs to be modified for most actual maps.)

Abandoning lines of force, let us glance at a remarkable episode in the history of mathematics. In the early nineteenth century, mathematicians finally realized, after centuries of believing that traditional Euclidean geometry had to be true, and the only possible or imaginable geometry, that it was not. They discovered that other kinds of geometry 'made sense'. One of these *non-Euclidean* geometries, ironically, was much like the spherical geometry which had been studied since the Greeks, because it was needed by navigators and geographers. In traditional spherical geometry, straight lines are arcs of great circles. One of the curiosities of this geometry is that the angle sum of a triangle is always greater than π (radians). In addition, you can calculate the area of a triangle from the angle sum. If the radius of the sphere is 1, so that the total surface area is $4\pi r^2 = 4\pi$, and the angles are A, B and C, then the area is $A + B + C - \pi$. In Fig. 7.28, the triangle is actually formed by the equator and two lines of longitude separated by $\frac{1}{2}\pi$, so its angles are all right-angles, or $\frac{1}{2}\pi$, and $\frac{1}{2}\pi + \frac{1}{2}\pi + \frac{1}{2}\pi - \pi = \frac{1}{2}\pi$. This is correct, because this triangle is one eighth of the surface of the whole sphere.

Another kind of non-Euclidean geometry can be represented inside a circle. A straight line is now an arc of a circle which cuts the given circle at right angles (Fig. 7.29). These include all the diameters of the

Fig. 7.28

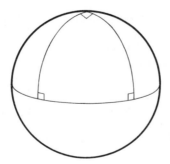

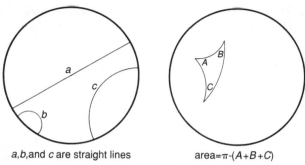

a,b,and c are straight lines   area=π-(A+B+C)

Fig. 7.29

circle. A triangle now has angles whose sum is *less than* 180°, and the area of any triangle is equal to $\pi - (A + B + C)$, the very reverse of the formula for spherical geometry. This kind of geometry is called *hyperbolic*, and the (earlier) non-Euclidean geometry on the sphere is called *elliptic*. That leaves – surely – some kind of *parabolic* geometry to complete the analogy to which we are by now quite accustomed.

What is parabolic geometry? Well, if we take a sphere and make it very large indeed, quite gigantic, and then look at only a small portion of its surface, that surface will appear very flat, and Euclidean geometry will be true to a very close approximation. If we allow the sphere to become infinite in size, then we actually get Euclidean geometry. On the other hand, if we take the hyperbolic geometry inside the circle, and allow the circle to expand infinitely, then hyperbolic geometry also becomes Euclidean, provided that we only look at a tiny portion of it. So traditional Euclidean geometry appears as the 'parabolic' special case lying, as expected, between elliptic and hyperbolic geometries.

Solutions to problems, Chapter 7

7A: *Searching for the widest angle*

At every point on the larger arc of the circle, the angle subtended by *AB* is the same, by the 'angle in the same segment' property, while it is greater from points inside the circle and less from points outside. (So the circle can be thought of as the boundary curve between the set of points from which *AB* appears smaller and the set of points from which *AB* appears larger – one typically mathematical way of looking at a curve.)

7B: *How many matches, moves or breaks?*

The answer in each case is 'It makes no difference'. 35 tennis players will be reduced to one winner in 34 matches, no matter how they are arranged. The 500 pieces of the puzzle will be reduced by 1 with each move made, and the number of pieces of chocolate will increase by 1 with each piece broken.

7C: *Bouncing on water*

Yes, there is. Rays of light passing through air and striking water at an angle will be partly reflected and partly refracted: the shallower the angle at which they strike the water, the more light is reflected, which is why you see such vivid reflections of the distant shore in the waters of a lake. If the rays of light are travelling through water, the denser medium, and strike the water–air surface at a sufficiently shallow angle, called the *critical angle*, then they will be totally reflected.

If a hard object strikes the surface of water at a shallow angle, it can actually bounce. Children – and adults! – play at skimming stones with flat sides which may bounce several times before their speed drops and they sink. In the days of cannons, it occasionally happened that a cannonball fired over water in a naval display would bounce on the surface and injure spectators on the far side of the harbour, as Descartes noted.

7D: *Rabbi Ben Moses' square vessel*

The analogy is exact. If the side of the square base is s and the height of the vessel is h, then

$$\text{total area} = 4sh + s^2 = 2sh + 2sh + s^2,$$
$$\text{volume} = s^2h = \tfrac{1}{2}\sqrt{(2sh \times 2sh \times s^2)}.$$

If we take the total area as constant, then $4V^2 = 2sh \times 2sh \times s^2$ is the product of three quantities whose sum is constant, and is therefore a maximum when they are all equal, which is when $h = \tfrac{1}{2}s$. (See also Problem 12F.)

7E: *The ballot problem*

If we represent the counting in a ballot by a point moving left to right, moving up one step whenever P gets a vote, and down one step whenever Q receives a vote, then the result is a random walk which, however, must not touch (let alone cross) the axis since P is always ahead of Q.

The point about brackets in any algebraic expression, as you move through the expression from left to right, is that the number of left-hand brackets is always greater than or equal to the number of right-hand brackets, simply because a bracket is opened before it can be closed. Therefore, any bracket sequence is also a possible sequence of votes in the ballot problem, with the slight difference that the excess of 'votes' cast for the 'left' candidate is always positive or zero, rather than positive. Every allowable bracket sequence can also be drawn as a random walk which may touch the axis, any number of times, but never crosses it, and will reach it again at the end.

7F: *Counting random walks*

Suppose that a random walk from the origin to (x,y) consists of p steps up and q steps down. Then $x = p + q$ and $y = p - q$. Each random walk consists of a distinct way of selecting the p positions for the up steps, since the remaining positions then have to be occupied by the down steps. Therefore to count the number of random walks is equivalent to counting the number of ways of selecting p positions from $p + q$ in a line. This happens to be

$$(p + q)(p + q - 1)(p + q - 2) \cdots (q + 1)(q)/p! = (p + q)!/p!q!.$$

7G: *Constructing a tangent to an ellipse*

As in Fig. 7.30, join A and B to P and bisect the angle $\widehat{APB}$, externally. This is the tangent to the invisible ellipse at P. The construction can be performed by extending one of the lines beyond P, as in the second figure, marking two points X and Y, equidistant from P, and then using the compass to draw two arcs, centred on X and Y, with the same radius, meeting at Z. Then PZ is the external bisector.

Fig. 7.30

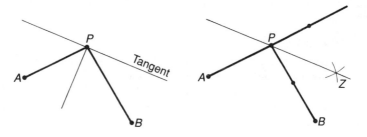

7H: *Tangents, foci and the directrix*

Yes, there are. The tangent to the hyperbola bisects the angle between *AP* and *PB* internally, whereas with the ellipse the angle is bisected externally. The tangent to the parabola bisects the equal lines to the focus and the directrix.

7I: *Ellipses and hyperbolas with the same foci*

By drawing, either by hand or with a computer, you will find that any ellipse and any hyperbola, provided that they share the same two foci, will be orthogonal, meaning that they intersect at right angles. Figure 7.31 shows two sets of orthogonal ellipses and hyperbolas. (There are also pairs of sets of parabolas, each parabola of one set being perpendicular to every parabola of the second set.)

Fig. 7.31

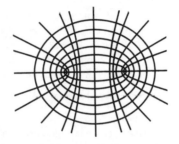

7J: *Plot the flow*

The flow will look something like Fig. 7.32.

Fig. 7.32

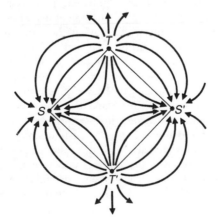

Chapter 8
Certainty, proof and illumination

Mathematicians have always been very confident, even cocky, about the certainty of their results. They can point to long sequences of theorems that have stood the test of time, unlike early scientific theories which have almost all been discarded as inadequate. The theorems in Euclid, or Archimedes, or Apollonius are as valid today as they were two thousand years ago. The science of Aristotle is no more than a museum piece. How certain is mathematics? How confident ought mathematicians to be?

Mathematicians have a big advantage even when they are behaving like scientists and doing experiments. Their experiments are easy to perform, and give great confidence, because the mathematicians are, as it were, *in control*. They are completely confident that they know what a whole number is, and that they can recognize a prime number; so, if they experiment to try to find the solution to Problem 8A, for example, they do not expect any nasty surprises. Physicists, in contrast, may *believe* that they know what an electron is, but they can never escape the possibility that electrons do in fact have an unknown property, which may be interfering with their particular experiment. So far, apparently, so good, and yet we have already seen in Chapter 3 that there was a time when mathematicians were confident that they knew what a 'polyhedron was', and went ahead and proved Euler's relationship by fundamentally flawed arguments.

Do mathematicians know exactly what they are talking about, or don't they? This is the first bridge that mathematicians have to cross in their search for certainty, and it has to be admitted that they cannot cross it with complete confidence, so their search for absolute certainty is already compromised. They can certainly rely on the long history of their most basic ideas, and the universal agreement that these concepts 'work'. However, this is another way of saying that, if a concept is new and untried, then even mathematicians should be wary of it. They may well discover with experience that their intuitive expectations of it are false. Indeed, this is one of the great values of experimental mathematics – experiments provide the mathematician with data against which mathematicians' natural expectations can be continually tested.

Problem 8A   *The converse of Fermat's Little Theorem*

We saw in Chapter 2 that, if p is prime, then $2^{p-1} - 1$ is divisible by p. For example, $2^6 - 1 = 63$ is divisible by 7. The ancient Chinese thought that, *if $2^{p-1} - 1$ is divisible by p, then p must be prime*. So did the great philosopher and mathematician Leibniz. They were wrong, because $2^{340} - 1$ is divisible by 341, although $341 = 11 \times 31$ is not prime. Prove that $2^{340} - 1$ is divisible by 341.

Problem 8B   *The coin-tossing game and random walks*

We introduced the idea of a random walk, and the zigzag graph that shows its progress, in Chapter 7. For this test, please imagine that you are playing a coin-tossing game with a partner, in which you score for heads and he or she scores for tails. The number of tosses is decided in advance. Test your intuitive feel for probability by assessing whether the following statements are true or false.

1. In a long coin-tossing game, each player is likely to be ahead for about half the time in total.

2. The lead can be expected to change quite frequently from one player to the other.

3. There is a 50–50 chance that one player will be in the lead for the whole of the second half of the game.

4. The lead is more likely to change towards the beginning of the game, rather than at the end.

5. In a game of 20 tosses, the probability that one player leads all the way is more than one third.

6. You play two games: if the second game is twice as long as the first, then the lead will change about twice as often.

7. In a game of any length, the probability that the lead never changes is greater than the probability that it changes once, which is greater than the probability that it changes twice, which is greater than the probability that it changes three times, and so on.

8. In a game of 10,000 tosses, the probability that one player is in the lead for more than 9900 steps, and the other in the lead for less than 100, is greater than 1 in 10.

Our intuitions are seldom more mistaken than when estimating probabilities. Some of the questions of Problem 8B have been wrongly answered even by professional statisticians. Proving the correct answers, by reasoning and calculation, provides some surprising results, and so does a computer experiment with a simple program to simulate the results of tossing a coin many times. The effect of counterintuitive conclusions is then to force mathematicians to examine their old and mistaken ideas, and to replace them with deeper and clearer ideas. 'Proof,' as a great mathematician once said, 'is the hygiene of mathematics,' and it forces mathematicians to clear out the dirt and rubbish and clean their ideas up. It is a bonus that, as they do so, they are likely to think of all sorts of new and interesting questions.

Fortunately, mathematicians have another reason for being confident. As G. H. Hardy put it, mathematics is created in the head, in the mind, like the rules of an abstract game. The ideas for the rules may come from the real world (the rules of chess are based on early warfare) but they do not depend on their origins for their existence. The natural numbers, 1, 2, 3, 4, 5, . . . , are constructed according to very simple rules. Anyone who can understand the rules of draughts or chess can understand how the pattern of Arabic counting continues. Because they are so simple and so game-like, they are extraordinarily widely shared, and we can say that the rules of ordinary arithmetic are tested by hundreds of millions of people every day all over the world.

We can be just as confident in practice of the rules of elementary algebra, so anyone who understands these rules can check for themselves that the following expression on the left has the two factors on the right:

$$x^5 + x + 6 = (x^2 + x + 2)(x^3 - x^2 - x + 3).$$

A rather more complicated sequence of moves demonstrates with only slightly less certainty that the polynomial

$$x^5 - x + N$$

will factorize into a quadratic and cubic, but cannot be factorized further, if and only if $N = \pm 15$, $\pm 22{,}440$, or $\pm 2{,}759{,}640$. (For example, if $N = 2{,}759{,}640$, then the factors are $x^2 + 12x + 377$ and $x^3 - 12x^2 - 233x + 7320$.)

Such arguments are just as convincing as simple analyses at chess or go. However, at this point, once again, we are brought up short by

Problem 8C   *A magic square*

8	1	6
3	5	7
4	9	2

Fig. 8.1

Figure 8.1 shows a 3-by-3 square divided into nine cells. The rules of the puzzle are simple: place the numbers 1 to 9, one in each cell, so that each row and each column, and both the diagonals, sum to the same total. This is easily done by trial and error. One solution is in the right-hand figure, and this square can be rotated or turned over to produce seven more solutions, a total of eight. No amount of trial and error will produce any more; but, of course, mere trial and error will not prove that there are none.

How might intuition help the process of trial and error? How can logical argument be used to find the solution much more quickly than trial and error, and to prove that no other solutions exist?

a practical possibility: what if the analysis is not so brief? It is a notorious fact that analyses at chess are very often flawed, not because the rules are contradictory, but because the analyst overlooks possible moves. How can we draw a line, in chess or game-like mathematics, between short analyses which are completely reliable and longer analyses which may contain flaws? Once again, we cannot: it has to be admitted that the longer a mathematical proof, the more likely it is that something has been overlooked.

A book *Erreurs de Mathematiciens des Origines a Nos Jours*, published in 1935, listed errors by 355 mathematicians, including Descartes, Fermat, Leibniz, Newton, Euler, Gauss, Poincaré and many other famous names. Many published proofs, which have been examined carefully by the author, and the journal editor, neither of whom wish to make fools of themselves, not infrequently contain errors. Sometimes these errors can be corrected, sometimes they prove fatal.

The more famous a problem, the more mathematicians are likely to try their hand at it, and fail. The very first book on number theory to be published in this country was by Peter Barlow, more famous for his *New Mathematical Tables*, who produced a false proof of Fermat's Last Theorem. We have already seen that Lamé's attempt to prove Fermat's Last Theorem was deeply flawed. His embarrassment was increased when a colleague, believing that Lamé had succeeded, proposed that a subscription be raised for 'the greatest mathematical discovery of the century'. Subsequently, many false proofs were offered, especially after Paul Wolfskehl in 1908 bequeathed to the Göttingen Academy of Sciences a prize of one hundred thousand marks for a satisfactory proof. More than a thousand false proofs were published in the following three years! It is perhaps fortunate that Wolfskehl's prize was hopelessly devalued by the German inflation after the First World War.

More recently, a Japanese mathematician, Matzumoto, claimed to have proved the *Riemann hypothesis*. This, not Fermat's Last Theorem, has been regarded by most professional mathematicians as the greatest unsolved problem of the present age, so news of Matzumoto's achievement even reached the popular press. It was then agreed, by mathematicians who had studied his arguments, that they were flawed, and Riemann's hypothesis returned to the 'unsolved' shelf.

It is understandable that false proofs of a very difficult problem should be offered, if regrettable that their number should vary with the financial reward on offer. However, mathematical arguments do not have to be very long to show errors. Cauchy was the first mathematician to write at length on the theory of groups. A group is nowadays defined very simply and clearly, so we might imagine that his work was accurate. Far from it! His results are full of errors. For example, Cauchy defined a special kind of group called a *substitution group*, and tried to list all the substitution groups with six or fewer members, but his understanding of substitution groups was defective, leading to false results.

Cayley also wrote extensively on groups. With Cauchy's example behind him, Cayley might have done better, yet his works contain numerous errors. Thus in *The American Journal of Mathematics* he stated that there were three groups of order 6, giving two sound examples and a third example which was just a duplicate.

More recently, group theorists have tackled the problem of completely classifying all examples of a special kind of group, called the *simple* groups. Many simple groups are easy to list, but a few are very difficult. These are the *sporadic* simple groups. The sporadic groups are few and far between, and difficult to find, which makes their capture a challenge. They are also bizarre, not least because some of them are so large. The largest, named by group theorists the *Monster*, is one of the largest mathematical objects ever considered: it has

$$808,017,424,794,512,875,886,459,904,961,710,757,005,754,368,000,000,000$$

elements. The complete proof of the classification of all the simple groups is also very large: it runs to more than 10,000 pages in papers scattered between hundreds of journals. Some group theorists naturally suspect that it contains gaps.

How can errors in mathematical proofs be avoided? One solution, if the process can be described precisely, and turned into a program, might be to use a computer. Surely, computers do not make mistakes in calculation? Computers have actually been used to 'solve' a very famous problem, the four-colour conjecture. The problem first occurred to Francis Guthrie while he was colouring a map of England. Guthrie's brother told Augustus de Morgan in 1852, but not until 1878 did it appear in print, in a question from Cayley in the *Proceedings of the London Mathematical Society*, who asked if the conjecture had been proved. It had not, but this public query started the conjecture's career, which bears a strong resemblance to the careers of Euler's relationship and Fermat's Last Theorem, in two respects: all of the early 'proofs' were erroneous, but many fruitful new ideas were developed in the attempts to prove the conjecture.

To cut a long story short, the conjecture was finally proved in 1976 by Kenneth Appel and Wolfgang Haken. Or was it? Haken and Appel's argument involved complicated initial analysis, followed by the use of 1200 hours of computer time, and 10^{10} separate operations, to analyse 1936 special cases. The computer finally said yes, but many mathematicians were unimpressed. They wanted to be able to examine the proof themselves, at leisure, and in detail, as they would any ordinary proof, to assure themselves that it was sound – most mathematicians are too independent to take anyone else's word for it! Yet they could no more examine the computer proof in detail than they could examine all 10,000 pages proving the classification of the simple groups.

Some mathematicians were even more dismissive. Paul Halmos asked the pertinent question:

> What did we learn from the proof? What do we know now that we didn't know before? . . . To be sure: I am not going to spend my time looking for a counterexample to the four-colour assertion. The printout had at least that

practical effect: it discouraged attempts to prove it wrong. Except for that, however, I feel that we, humanity, learned mighty little from the proof; I am almost tempted to say that as mathematicians we learned nothing at all. Oracles are not helpful mathematical tools.

He then revealed what he, and many other mathematicians, really desire: 'We need a simple insight into a new and complicated kind of geometry or intricate algebra . . .' Having discovered such a new algebra, or geometry, the proof of the four-colour conjecture would then ideally appear as the conclusion of an argument, using these new ideas, which would be short enough for a competent expert to grasp, as a whole.

Halmos believes that 'mathematics is, after all, not a collection of theorems, but a collection of ideas.' Given the right ideas, every argument can be expressed in meaningful thought-sized chunks, and neither 10,000 journal pages nor 1200 hours of computer time will be necessary. As Richard Hamming, the inventor of the Hamming code, put it, 'Some people believe that a theorem is proved when a logically correct proof is given; but some people believe that a theorem is proved *only* when the student sees why it is inevitably true. The author tends to belong to this second school of thought.'

Mathematicians have to rely on game-like proofs, consisting of a sequence of logical steps, which are often extremely elegant and powerful, and easily grasped as a whole, like Fermat's proof that two squares cannot also be the sum and difference of squares. Yet there is a special attraction in the proof that relies as little as possible on a chain of argument, and as much as possible on a *way of looking*, a special perception, which when discovered reveals the solution to be totally obvious, and totally convincing. 'Ah! *now* I see it! Of course! How clear!' thinks the delighted mathematician, who only a moment ago did not find it obvious at all. Such proofs provide *illumination*. They do not merely convince, more or less, by experiment; they do not convince by a chain of argument, but by making clear a deeper meaning within the problem.

In this book, we have already seen examples of how triangular patterns of dots fit together to make rectangles or squares, or how the hexagonal numbers can be seen as fitting together, like so many shells, to make a cube. Such pictures make algebraic proofs redundant. At a higher level of complexity, the inferences that we drew from the theorem on page 178, leading all the way to Napoleon's theorem, were based on different ways of looking at the original theorem (at

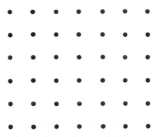

Fig. 8.2

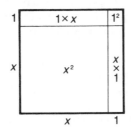

Fig. 8.3

special cases in particular), rather than on algebra, though all those results can be proved by algebraic moves.

The simplest illumination comes from pictures that school pupils can appreciate. The dot picture in Fig. 8.2 shows 6 rows of 7 each which is the same number in total as 7 rows of 6 each, by turning the page round. Figure 8.3 makes clear that $(x + 1)^2 = x^2 + 2x + 1$. Figure 8.4 makes fairly clear that $(x + 1)^3 = x^3 + 3x^2 + 3x + 1$, though it must be admitted that the degree of clarity is now falling somewhat (the third slab representing x^2 is on the back face). The geometric picture approach stops at three dimensions.

Fig. 8.4

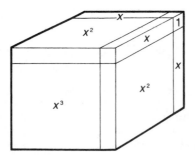

Problem 8D *Computer testing for Mersenne primes*

Desk calculators, followed by electronic computers, have long been used to test certain special numbers to see if they are prime. The simplest test works on Mersenne numbers, named after Marin Mersenne, a philosopher, theologian, mathematician and musician, who was a friend of Descartes, Desargues, Fermat and Pascal, and who corresponded with all the mathematicians of Europe, acting as a sort of clearing house for problems and ideas. In 1644 he gave a list of prime numbers p for which, so he claimed, $2^p - 1$ is prime. His list contained errors, but prompted mathematicians to further study numbers of this special form. In the nineteenth century, Eduard Lucas, who invented the well-known Tower of Hanoi puzzle and other mathematical recreations, discovered this test, to decide whether a Mersenne number is prime.

Suppose that we want to know if the fifth Mersenne number, $2^5 - 1 = 31$, is prime. We start with the number 4 (one less than 5), and we repeatedly square it, and subtract 2. At each step, however, we reduce the result by taking only the remainder when it is divided by 31. So our test goes like this:

$$n_1 = 4,$$
$$n_2 = 4^2 - 2 = 14,$$
$$n_3 = 14^2 - 2 = 194 = 6 \times 31 + 8,$$
$$n_4 = 8^2 - 2 = 62 = 2 \times 31 + 0.$$

There is no remainder at the *fourth* step, and so the *fifth* Mersenne number is indeed prime. If there had been a remainder, it would have been composite.

For readers who have a computer: is the 13th Mersenne number, 8191, prime or composite?

The pattern in the purely algebraic argument might seem at least as convincing:

$$(x + 1)^3 = (x + 1)(x + 1)^2 = (x + 1)(x^2 + 2x + 1)$$

and, when we multiply out the last bracket, we not only get the usual result, but we can see also, what is not obvious from the geometrical model, that the middle coefficients of $(x + 1)^3$ are obtained by adding

Problem 8E   *Proof by looking*

These two puzzles are variations on the same theme:

1. How can a rook move from one corner of a chessboard to the diagonally opposite corner, passing through each square exactly once?

2. How can a gaoler enter a square block of 16 cells at one corner, and leave through the diagonally opposite cell, having entered each of the 16 cells exactly once?

What is the answer in each case? How can the answer be made quite obvious by suitably changing each diagram?

Problem 8F   *More proofs by looking*

Fig. 8.5

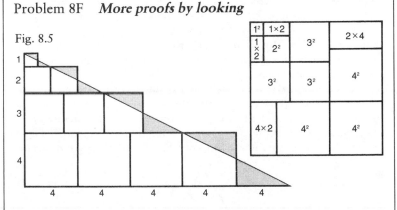

1. How do the upper and lower diagrams in Fig. 8.5 each show the sum of the series

$$1^3 + 2^3 + 3^3 + 4^3 + 5^3 + \cdots \ ?$$

2. Areas can be subtracted from each other, as well as added. With this possibility in mind, find a figure which shows that

$$1^2 = 1,$$
$$2^2 - 1^2 = 3,$$
$$3^2 - 2^2 + 1^2 = 6,$$
$$4^2 - 3^2 + 2^2 - 1^2 = 10,$$

and so on. The numbers on the right are the triangular numbers.

the coefficients of $(x + 1)^2$ in adjacent pairs, which then leads to Pascal's triangle which we have already seen:

$$
\begin{array}{ccccccc}
 & & & 1 & & & \\
 & & 1 & & 1 & & \\
 & 1 & & 2 & & 1 & \\
1 & & 3 & & 3 & & 1 \\
\end{array}
$$

. . .

As always, 'perception' in mathematics does *not* mean just the visual perception required to look at a geometrical figure, but also applies to algebra and arguments in symbols.

Many problems – perhaps all – can be proved by a variety of methods, including algebraic transformations and novel ways of looking. We conclude this chapter with three examples of problems that can be tackled by scientific experiment, by game-like argument, and by changes in perception, in order to illustrate the differences between these approaches.

The first is a puzzle which asks you simply to count the number of triangles in Fig. 8.6. Not difficult, but it would be easy to make a mistake, missing a triangle or counting one twice, so it is tempting to try some simpler versions of the same figure first, as in Fig. 8.7. These are much simpler to count with confidence, and the pattern seems

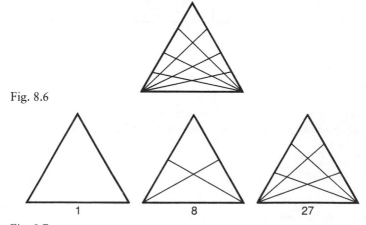

Fig. 8.6

Fig. 8.7

obvious. The number of triangles in the original figure will be $4^3 = 64$. This is the end of our simple scientific investigation. We feel confident, but we are also aware of the kinds of misleading series we saw in Chapter 3, so our confidence is not unbounded.

To bolster our confidence further, we decide to calculate. To make the most of the calculation, we take a general case, in which $n - 1$ lines through each base vertex divide the opposite sides into n parts each. A helpful step is to remove the base of the triangle (as on the left of Fig. 8.8) to count the triangles which do not include it, and then to replace it (as on the right) to count those that do. In the left-hand figure, a triangle must include two sides from the left vertex, which can be chosen in $\frac{1}{2}n(n - 1)$ ways, and one side from the right vertex, which can be chosen in n ways – or the other way round. The total number of choices is double $n \times \frac{1}{2}n(n - 1)$ or $n^2(n - 1)$. In the right-hand figure, we have already chosen the base as one side. The other sides must come one from each vertex, so each can be chosen in n ways, and the total number of choices is n^2. The grand total of triangles is $n^2(n - 1) + n^2 = n^3$.

Fig. 8.8

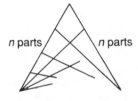 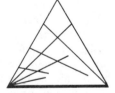

n parts   n parts

Problem 8G   *A variant problem*

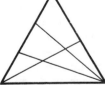 

Fig. 8.9

The number of divisions along the triangle's sides do not have to be equal. In the diagrams of Fig. 8.9, there is a different number of parts on each side – say m and n. How many triangles are there?

Problem 8H *Counting rectangles in a rectangle*

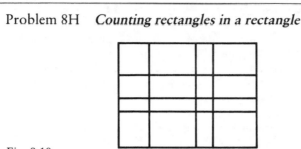

Fig. 8.10

The rectangle shown in Fig. 8.10 has been divided by n vertical and n horizontal lines. How many rectangles does it contain in total? What is the connection between the number of rectangles and the triangular numbers?

This is just as we supposed, and could be the end of our investigation. However, this correct solution still leaves an obvious question unanswered. Why is the solution *so simple*? Why is the formula not more complicated? Coincidence? Mathematicians are not inclined to believe in coincidences. Figure 8.11, a flattened-out version of Fig. 8.6, gives a slight three-dimensional effect. Since n^3 is the number of unit cubes in an n by n by n cube, could there be some correspondence between the triangles and the unit cubes? Taken together, the simplicity and visual illusion suggest there is more to discover. Here is one route which goes further into the problem.

In Fig. 8.12, we have broken the base of the triangles and moved the lines proceeding from each vertex, slightly away along the broken sides of the base. This, of course, is not the kind of step that you take if you are using algebra, but it is typical of how a problem can be looked at in a novel way. There are no longer any triangles at all, but every triangle in the original figure (Fig. 8.6) is now a quadrilateral with either one or two sides actually along the broken base. If we continue the same transformation of the original triangle, the original triangle eventually becomes a square, and each quadrilateral becomes a rectangle, with one or two sides actually along the bottom edges of the square (Fig. 8.13). This visual change of perspective has turned the original puzzle into a different one, which we can solve if we know how many such rectangles are formed by the lines of a square divided into unit squares.

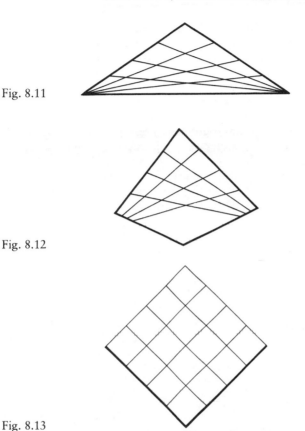

Fig. 8.11

Fig. 8.12

Fig. 8.13

By inspection of Fig. 8.13, the number of rectangles with at least one edge along the two bottom edges of the outer square is the total number in the complete square, less the number in the inner (3 × 3) square: so the total is $T_n^2 - T_{n-1}^2$, which we know from page 46 is just n^3. Well, well! It appears that the three-dimensional quality of the original figure is *probably* itself an illusion. So is the simplicity of n^3. The puzzle seems to be really about triangular numbers. Or is it? This is certainly one way of looking at the original puzzle, but we cannot eliminate the possibility that there is another way of looking at it which *will* reveal a direct connection with the division of a large cube into unit cubes.

*

Pythagoras' theorem is plausibly the oldest geometrical theorem in the world. Pythagoras lived *c.* 540 BC and founded a school of philosophy which taught that numbers – that is, the integers – are the basis of the universe. It seems that the Greek interest in triangular and square numbers, and the simple patterns that express them, originated with Pythagoras. He is also credited with the independent discovery of the geometrical theorem that bears his name: in the time-honoured formula, 'in any right-angled triangle, the square on the hypotenuse is equal to the sum of the squares on the other two sides.' (See Fig. 8.14.) In other words $AC^2 = AB^2 + BC^2$, where B is the right-angled corner and AC is the hypotenuse. Whether or not Pythagoras discovered this himself, the theorem was known to the Babylonians in the time of Hammurabi the law-giver, and it has been conjectured that a form of Pythagoras' theorem was known to neolithic man in Europe. It appears in China in a manuscript dated between 500–200 BC, the *Chou Pei Suan Ching*, or 'Arithmetic Classic of the Gnomon and the Circular Paths of Heaven'.

Because it is so old, famous and important, and perhaps for other reasons, several hundred proofs have been published. E. S. Loomis presented 367 of them in his book, *The Pythagorean Proposition*, published in 1940, and no doubt more proofs have been discovered since. Which are the most convincing? And which are the most illuminating?

Fig. 8.14

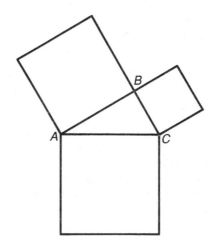

Problem 8I   *Leonardo's proof of Pythagoras' theorem*

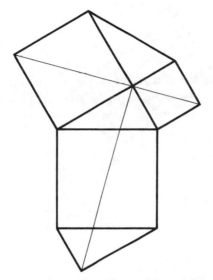

Fig. 8.15

Leonardo da Vinci took a standard figure, added a copy of the original triangle to the opposite side of the large square, and joined the corners of the smaller squares, as shown in Fig. 8.15, to make a further copy of the triangle. He then added the thin lines shown in the figure. How does this help to prove Pythagoras' theorem?

The crudest approach is simply to draw an accurate diagram and measure the lengths of the sides. Crude, but effective, and after a few such experiments anyone is likely to believe that Pythagoras' theorem is true. Yet the reasons for its being true remain a total mystery.

By taking a special case, such as the tiling pattern on the left of Fig. 8.16, it is easily seen that it sometimes works. The tiling pattern on the right rather suggests that it always works, at least to a close approximation: what happens to the jagged edge of the large 'square'? It also perhaps hints at a reason why, but this reason remains obscure.

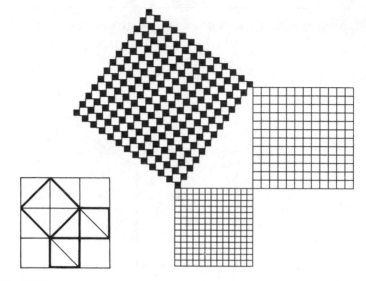

Fig. 8.16

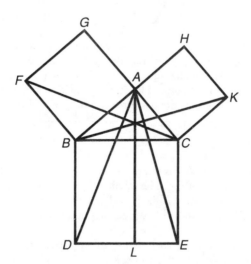

Fig. 8.17

Euclid proved Pythagoras' theorem in his *Elements*, as Proposition 47 of Book 1, using the construction shown in Fig. 8.17. He started by proving that triangles *ABD* and *FBC* are identical, and so are the pair *KCB* and *ACE*. After some more argument, he concludes that the

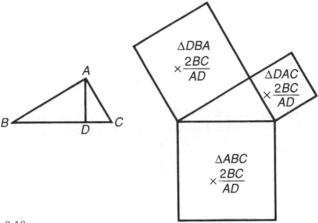

Fig. 8.18

square $ACKH$ is equal in area to the rectangle whose diameter is CL, and similarly square $ABFG$ equals the rectangle on BL as diameter, and the conclusion follows.

This proof is completely convincing, in the way that all such proofs in elementary geometry are. However, it does nothing to *illuminate* the proposition. Why should Pythagoras' theorem exist in the first place? For that matter, why is it so important? It suggests that Pythagoras has something to do with congruent triangles, but then most simple theorems in Euclid are proved with the aid of congruent triangles, and Fig. 8.18 proves Pythagoras much more simply by using *similar* triangles. In the left-hand figure, $\widehat{BAC}$ is a right angle, and AD is an altitude, dividing ABC into two smaller right-angled triangles. Area ABC = area ABD + area ACD. But the squares in the right-hand figure are proportional in area to the triangles in the first figure, because each square is constructed on the hypotenuse of the corresponding triangle, and so $BC^2 = AB^2 + BC^2$.

Problem 8J   *Extra properties in Euclid's figure*

Euclid's figure has a number of interesting properties that are not relevant to the theorem. How many can you discover?

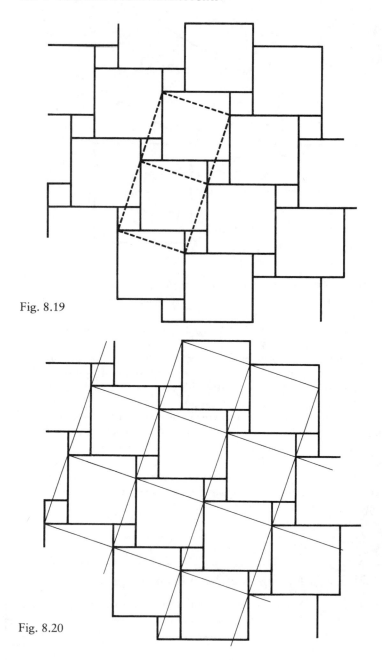

Fig. 8.19

Fig. 8.20

There is a clue to the deeper meaning of Pythagoras in the tessellation shown in Fig. 8.19. This tessellation can be obtained from a chessboard tessellation of identical squares, by shifting them apart in two directions. By varying the amount of movement, squares of any two sizes (as well as different sizes of rectangles) can tessellate the plane, repeating for ever. This repetition is regular, and we can join corresponding points, such as the top left corners of the large squares, to form another, simpler, tessellation (Fig. 8.20). Each thin-line square is the square on the hypotenuse of a right-angled triangle whose other sides are the sides of the two original sizes of squares.

The double tessellation shown in Fig. 8.20 now proves Pythagoras' theorem: either we can say, 'over the whole plane, there is one large square (forming the thin grid) for each pair of medium-and-small squares, so the large square equals to the sum of the other two in area'; or we can actually pick out the dissection which will cut the large square into five pieces which fit together to make both the smaller squares (Fig. 8.21).

This proof is far, far simpler than Euclid's. It is so simple, in fact, that we might find it suspicious: we seem to have proved a theorem that in Euclid takes a couple of dozen lines and a string of arguments, by just looking at a picture. Have we missed something out? What have we assumed surreptitiously, without acknowledgement?

We made at least two basic assumptions: firstly, that it is possible to tile the plane with identical squares in a chessboard pattern; secondly, that it is possible to slide them over the plane without distorting them in any way, to create a pattern of identical small squares in between them. Now we are getting somewhere: Pythagoras depends directly on some very simple and basic facts about the *surface that the figure is drawn on* – and that is why it is so important.

If we can tile that surface with identical squares, and slide the squares about in the manner described, then Pythagoras' theorem will be true. If we cannot, then we can plausibly assume that Pythagoras will be untrue. This is so. On a circular cylinder, we can arrange and move squares as described (because a cylinder can be cut and

Fig. 8.21

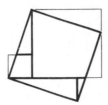

Problem 8K   *More dissections of two squares*

What other dissections of the large square into the smaller squares would be obtained, if instead of joining the top left corners of the medium squares to form a new network, the centres of the smaller squares had been joined?

What other dissections can be obtained by choosing different sets of corresponding points?

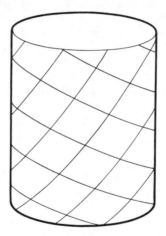

Fig. 8.22

opened out to make a portion of a plane) – see Fig. 8.22 – and Pythagoras works on a cylinder, provided that a straight line is defined to be, as usual, the shortest distance between two points across the curved surface of the cylinder. A sphere, however, cannot be tessellated with squares, and so Pythagoras does not work on the surface of a sphere.

We have already seen (on page 40) a simple proof of Fibonacci's formula for the product of two sums of two squares as a sum of two squares, generally in two ways:

$$(x^2 + y^2)(p^2 + q^2) = (xp + yq)^2 + (xq - py)^2$$
$$= (xp - yq)^2 + (xq + py)^2.$$

What could be clearer? Nothing, in terms of algebraic manipulation. This is perfectly clear. And yet, if we ask 'Why does this pattern *exist*

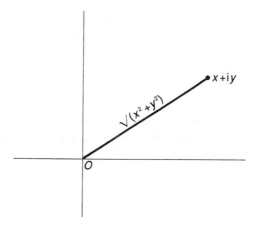

Fig. 8.23

in the first place?', then we have no answer. Is it perhaps a kind of coincidence? When you expand $(x + y)^2$ and $(x - y)^2$, the terms are identical apart from the differing signs of the $2xy$ term in the middle. It is just these differing signs which allow the middle terms $2pqxy$ and $-2pqxy$ to cancel out, so perhaps that's it – a neat trick, hiding no deeper meaning behind it.

This would have been a pardonable conclusion to any mathematician prior to the understanding of complex numbers. Complex numbers, however, have a subtle connection with sums of two squares. We can represent a complex number $x + iy$ as a point, or as a vector, as in Fig. 8.23. As a point, $x + iy$ has a distance from the origin, which is, by Pythagoras, $\sqrt{(x^2 + y^2)}$. As a vector, it has a length, which is the same. How important are these facts? They are very important because, when two complex numbers are multiplied together, the 'length' of the product is the product of the lengths of the original complex numbers:

Numbers $(x + iy)(p + iq) = (xp - yq) + i(yp + xq)$

Squares of lengths $(x^2 + y^2)(p^2 + q^2) = (xp - yq)^2 + (yp + xq)^2.$

This must be significant! It is no longer plausible that the product-of-squares pattern is 'coincidental' when it appears, naturally and inevitably, in the products of complex numbers.

Does this lead us anywhere else? It can do, if we spot a double connection: in one direction to Euler's formula for the product of the sums of *four* squares, and in a different direction to the discovery of *quaternions* by William Rowan Hamilton in 1843. Euler's formula, which he gave in 1770, is shown in the box below.

Euler's formula

$$(a^2 + b^2 + c^2 + d^2)(p^2 + q^2 + r^2 + s^2)$$
$$= (-ap + bq + cr + ds)^2$$
$$+ (ar - bs + cp + dq)^2$$
$$+ (as + br - cq + dp)^2$$
$$+ (aq + bp + cs - dr)^2.$$

(A number of variants can be created by permutations and/or sign changes of the letters.) There is an obvious resemblance to the formula for two-square products, but once again there is no indication as to *why* this relationship exists in the first place, though it is perhaps a hint that there is no similar formula for the product of sums of the three squares.

Turning to our other connection, Hamilton (so the story goes) was standing on a bridge near Dublin, when the solution to a problem that had been troubling him for fifteen years came to him in a moment of inspiration. When multiplying two algebraic quantities together, it is *not* necessary that $A \times B$ should always equal $B \times A$. Of course they do in traditional algebra, but Hamilton was not thinking of traditional algebra: far from it, he was thinking of a kind of hypercomplex algebra in which each 'number' was composed of four components, three of which were like the 'imaginary' component of a complex number. Hamilton's new numbers looked like this:

$$x + iy + ju + kv,$$

where the rules for multiplying i, j and k are that $i^2 = j^2 = k^2 = -1$, and ij = k, jk = i and ki = j, *but also* ji = -k, kj = -i and ik = -j.

This inspiration of Hamilton's was the most extraordinary since the discovery of complex numbers. A fact that mathematicians had taken for granted since the beginning of time – that it makes no difference in which order you multiply two numbers – was thrown overboard. However, just like complex numbers three hundred and more years earlier, Hamilton's *quaternion algebra,* as he called it, worked. It not only worked, but it had many practical applications. It led Hamilton to invent the terms *scalar* and *vector,* and to apply his extraordinary new algebra to problems in mechanics and physics, as well as to pure mathematics.

Our interest, however, lies in its connection with Euler's formula, so let's pursue the analogy with complex numbers. The 'length' of a quaternion $x + iy + ju + kv$ ought to be, by analogy, $\sqrt{(x^2 + y^2 + u^2 + v^2)}$. By the same analogy, we would expect (if quaternions really are like hypercomplex numbers) that the length of the product of two quaternions will be the product of their lengths. A small experiment is in order. We use the relationships between i, j and k to multiply two quaternions, and then we compare (1) the product of the squared lengths of the quaternions $a + ib + jc + kd$ and $p + iq + jr + ks$, and (2) the squared length of the product of these quaternions. The result is Euler's formula. (You may have to rearrange the terms slightly to get the exact layout of the formula given in the box above.)

If we now ask, 'Why do Fibonacci's two-square product and Euler's four-square products exist?' we can answer, 'Because of the existence of complex algebra and quaternion algebra.' At this point, you will of course spot a small snag: why do complex algebra and quaternion algebra exist? Why is there no algebra that would magic-ally produce a formula for the product of the sums of *three* squares? Algebraists have proved that such an algebra does not exist – but that is another story.

Solutions to problems, Chapter 8

8A: *The converse of Fermat's Little Theorem*

We have to show that $2^{340} - 1$ is divisible by 11 and by 31. By factorizing,

$$2^{340} - 1 = (2^5 - 1)(2^{335} + 2^{330} + 2^{325} + \cdots + 1),$$

so it is divisible by $31 = 2^5 - 1$. Similarly,

$$2^{340} - 1 = (2^{10} - 1)(2^{330} + 2^{320} + 2^{310} + \cdots + 1);$$

therefore it is divisible by $2^{10} - 1 = 1023 = 11 \times 93$, so it is divisible by 11.

Alternatively, we may list the remainders of the powers of 2 when divided by 11, in sequence as follows:

1	2	4	8	16	32	64	128	256	512	1024	...
1	2	4	8	5	10	9	7	3	6	1	2 4 ...

The sequence repeats from $2^{10} = 1024$ onwards. Therefore $2^{340} - 1$ is divisible by 11. Similarly, the remainders on division by 31 repeat in cycles of 5:

1	2	4	8	16	32	64	...
1	2	4	8	16	1	2	...

Therefore, $2^{340} - 1$ is also divisible by 31.

8B: *The coin-tossing game and random walks*

1. False. 2. False. 3. True. 4. True.

5. True: the probability that the lead never changes is about 0.352.

6. False: if the second game is twice the length of the first, the lead will only change about $\sqrt{2}$ times as often. If N times as long, the lead will change $\sqrt{N}$ times as often.

7. True.

8. True: in fact the probability that one player leads for more than 9930 trials, and other for fewer than 70, is greater than 0.1.

8C: *A magic square*

The commonest helpful intuition is that the middle number of those to be placed should go in the middle cell. In other words, the 5 goes in the centre. This is correct, and makes a trial and error solution far simpler.

The simplest logical step is to sum the nine digits, making 45, and then note that, since each row and column of three adds to the same total, this total must be $45/3 = 15$. There are then various arguments to deduce the exact positions of the numbers. Here is one sequence of steps.

Add up the row, column and two diagonals which pass through the centre cell. This total will be $4 \times 15 = 60$ and it includes every cell once, except the centre cell which is included four times. So the excess, $60 - 45 = 15$, is three times the centre cell, which must therefore be 5. It follows that the numbers sandwiching the centre cell must be pairs adding to 10. We therefore have to correctly place 1 and 9, 2 and 8, 3 and 7, and 4 and 6.

	9	
	5	
	1	

Fig. 8.24

If 9 goes in one corner, then three lines through that corner add to 15 and there are three pairs of numbers to complete those rows which add to 6. But this is impossible, since only 1 plus 5 and 2 plus 4 yield 6. Therefore 9 must go in the middle of one side. (Alternatively, we could have reasoned by a symmetrical argument that 1 must go in the middle of a side, because there are only two pairs of numbers to make a row containing 1 up to a total of 15.) Placing the 9 and the 1, say, as in Fig. 8.24, it follows that the corners must be the four even numbers, 2, 4, 6 and 8, and the first two sandwich the 9. The remaining cells can then be filled in only one way.

Since there are four possible middle-side positions for the 9, and the 2 and 6 can be placed on either side of the 9 in two ways, there are $4 \times 2 = 8$ possible solutions to the puzzle. In fact, these consist of any one solution and its seven reflections and rotations, so there is 'essentially' just one solution.

8D: *Computer testing for Mersenne primes*

8191 is prime.

8E: *Proof by looking*

1. It cannot. The opposite corners of a chessboard are the same colour, either both white or both black. It takes an even number of steps to go by rook moves from a square of one colour to another square of the same colour. But there are 63 squares, an odd number, to be passed through in order to solve the problem. The problem is therefore impossible. (The closest route to a solution will omit just one square.)

2. This problem is essentially identical to the first, as becomes clear if the cells of the gaol are coloured black and white like a chessboard. The gaoler cannot achieve this feat.

8F: *More proofs by looking*

In the first diagram of Fig. 8.5, the L-shaped borders contain 1^2, 2×2^2, 3×3^2, 4×4^2, ... The sides of the squares are 1, $2 \times 1\frac{1}{2}$,

3×2, $4 \times 2\frac{1}{2}$, ...: in other words, $\frac{1}{2}(1 \times 2)$, $\frac{1}{2}(2 \times 3)$, $\frac{1}{2}(3 \times 4)$, $\frac{1}{2}(4 \times 5)$, ... Therefore

$$1^3 + 2^3 + 3^3 + 4^3 + \cdots + N^3 = [\tfrac{1}{2}N(N + 1)]^2 = \tfrac{1}{4}N^2(N + 1)^2.$$

In the second diagram, each row contains N squares, each $N \times N$. When the diagonal line is drawn and the shaded pieces rearranged to complete the triangle, the base of the triangle is $N(N + 1)$ and its height is $1 + 2 + 3 + \cdots + N = \tfrac{1}{2}N(N + 1)$. Therefore, the area of the triangle $= \frac{1}{2} \times$ base $\times$ height $= \tfrac{1}{4}N^2(N + 1)^2$, as before. Figure 8.25 shows how to sum $5^2 - 4^2 + 3^2 - 2^2 + 1^2$, to leave three L shapes which make up the 5th triangular number.

Fig. 8.25

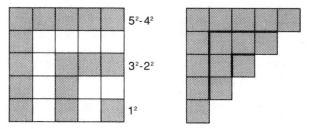

8G: *A variant problem*

If the sides of the triangle are divided into n and m parts, then there are mn triangles that include the base line, and $\tfrac{1}{2}m(m - 1)n$ triangles including two of the m lines on one side, and $\tfrac{1}{2}n(n - 1)m$ including a pair of the n lines on the other. This is a total of $\tfrac{1}{2}mn(m + n)$ triangles. This becomes the simple n^3 when $m = n$.

8H: *Counting rectangles in a rectangle*

Two sides of the rectangle will have to be chosen from the vertical lines, of which there are $n + 1$, including the right and left edges. Two lines can be chosen from $n + 1$ in $\tfrac{1}{2}n(n + 1)$ ways, or T_n ways, where T_n is the nth triangular number. Similarly, there are T_n ways of choosing a pair of horizontal lines for the other sides of the rectangle. So the total number of rectangles is T_n *squared*.

8I: *Leonardo's proof of Pythagoras' theorem*

The figure excluding the large square and the added triangle attached to it consists of two quadrilaterals, symmetrical about the line PQ. The original triangle, plus the large square, plus the added triangle

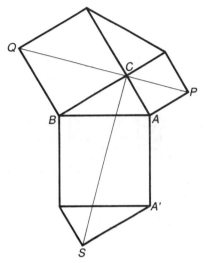

Fig. 8.26

also form two quadrilaterals, which are also congruent to each other, by symmetry about the middle point of the line CS. Moreover, if the quadrilateral $PQBA$ is rotated anticlockwise through a right angle, about the point A, it coincides with $CSA'A$. Therefore all the quadrilaterals are equal in area, and (after subtracting the triangles) the area of the two smaller squares equals the area of the larger.

8J: *Extra properties in Euclid's figure*

The lines AL, CF and BK concur and the pairs AE and BK, and AD and FC, are perpendicular.

8K: *More dissections of two squares*

It is a neat fact that *any* set of corresponding points can be joined to form a dissection of the large square into the two smaller ones. Figure

Fig. 8.27

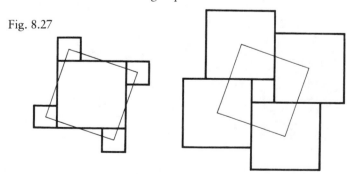

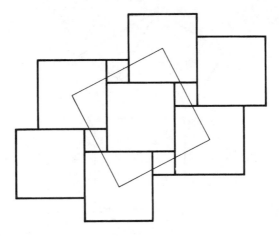

Fig. 8.28

8.27 shows two examples, using centres. Moreover, if corresponding points are chosen which are further apart, then dissections of two of the medium squares and two of the small squares into one square are produced, as in Fig. 8.28.

Chapter 9

Mathematics in science: searching for the truth

Mathematics undoubtedly originated in everyday activities: counting, measuring, sharing and exchanging. From these mundane but essential origins, mathematics continued to develop as a means of calculation, but at the same time it started to develop a life of its own. Mathematicians could not help noticing and pondering properties of numbers and shapes that had no necessary connection at all with everyday life. Many of these geometrical properties were collected together, and beautifully organized in a deductive system, by Euclid in his *Elements*, two hundred and fifty years after Pythagoras and his disciples were already speculating about properties of the whole numbers. Most of Euclid's propositions are of no practical use.

More than five hundred years later, Diophantos posed questions about integers that were puzzling and provoking, but practically useless. This was the puzzling, recreational source of modern mathematics. It is not to be sneered at. Far from it, the whole of modern number theory, large parts of algebra, much of geometry, and many

---

Problem 9A   *Problems from Diophantos*

Problems posed by Diophantos in his *Arithmetica*, requiring solutions in integers or fractions only, provided a major source of modern number theory, not least through the agency of Fermat, who studied and developed Diophantos' ideas. Here are two of his easy problems and a third which is much harder. The solutions are all integers.

1. Two numbers are such that if the first receives 30 from the second, they are in the ratio 2:1; but if the second receives 50 from the first, their ratio is then 1:3. What are the numbers?

2. The sums of four numbers, omitting each of the numbers in turn, are 22, 24, 27 and 20, respectively. What are the numbers?

3. Find three numbers such that the product of any two added to the third gives a square.

of the roots of graph theory and topology lay in just such puzzling questions.

Mathematics could have continued in this direction, developing as a result of the mathematician's individual curiosity, and sense of challenge. Indeed, in one part of the world, in Japan, that is exactly what did happen. The native, traditional Japanese mathematics, called *wasan*, was developed in almost (but not quite) complete isolation from Western mathematics, right up to 1865 when Japan was opened to the West by Commander Perry's expedition.

Wasan had also started as a utilitarian art, but later became an activity of leisured samurai, merchants and wealthy peasants who enjoyed *wasan* as much as they enjoyed writing *haiku*, or practising calligraphy or the tea ceremony. *Wasan* had its own great mathematicians, of whom the most renowned was Seki Kowa (*c.* 1642–1708) who studied equations with negative and imaginary roots, discovered the Bernoulli numbers before Jakob Bernoulli, and the concept of a determinant a decade before it was discovered in the West by Leibniz. *Wasan*, however, suffered from a serious disadvantage. Because it was treated entirely as an art, the purer the problem and the further it was from any practical application, the more highly it was regarded, and complexity for its own sake became a virtue.

Mathematics in the West also possessed an artistic side, but it had another stimulus, and another direction in which to grow. The Greek Archimedes (287–212 BC) was a sublime mathematician who solved problems about tangents, areas and volumes, by methods which were essentially those of the calculus, summed infinite series, solved a cubic by geometrical methods, created the sciences of mechanics and hydrostatics – he invented the Archimedean screw which is still used in the Middle East to raise water for irrigation – and wrote a treatise, the *Sand Reckoner*, in which he calculated the number of grains that would fill the universe, as the Greeks conceived it, devising his own system of numbers for the purpose. The works of Archimedes are a reminder that the title of Euclid's *Elements* was meant literally: they were the beginnings of Greek mathematics, not its end.

Starting with Archimedes, mathematicians and scientists persisted in asking questions with a mathematical flavour but in a scientific context. This source also was rich in consequences. The calculus and analysis, and much of topology and modern algebra, developed out of questions about the real world of business, science and technology. Archimedes analysed the working of levers and made the bold claim,

'Give me a place to stand on, and I will move the earth!' When asked by Hieron, the king of Syracuse, to demonstrate how a great weight could be moved by a small force, Archimedes loaded a three-masted ship with freight and crew, and drew it along by using his own strength multiplied by a compound pulley.

We have already seen how simple ideas of balances can be deduced, apparently, by common sense. If two equal weights are supported at the midpoint of a light rod joining them, then it is natural to suppose that they will balance (Fig. 9.1). We would be extraordinarily amazed, for example, if the weight on the left always fell and the weight on the right rose. (Of course, we are assuming that we have some way of deciding that the weights are equal in the first place, without using a balance; they might be made of equal volumes, say, of the same uniform material.)

What happens if we have 3 units at one end and 1 unit at the other? If we suppose that we can split the 3 units into three equally spaced units, then we could have the arrangement shown in Fig. 9.3.

Fig. 9.1

Problem 9B *Balancing a triangle*

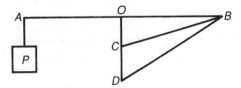

Fig. 9.2

In Fig. 9.2, the line AOB is a lever supported at its middle point, O, and BCD is a triangle, whose side CD is vertical and directly below O and attached to O by a thread. The vertex B is attached to the lever. Archimedes asserted that if an area P (meaning a rectangle of a certain size) is used to balance the weight of the triangle, then the area of P will be one third of the area of the triangle. Why is this conclusion justified?

Fig. 9.3

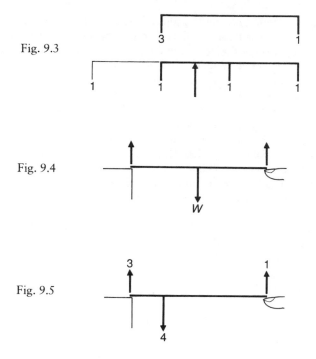

Fig. 9.4

Fig. 9.5

We now argue that the balance point, or centre of gravity, of the four equally spaced weights will be in their midpoint, and so we 'deduce' that the centre of gravity, or balance point, of the original unequal weights, is one quarter of the distance from the larger to the smaller.

In this way we can 'deduce' the law of the balance, by making only the very simplest and most plausible assumptions about how weights behave – and our deductions are amply confirmed by very simple experiments. In a similar manner, we can just as plausibly deduce the 'law of the lever'. In fact, we can simply turn the figures for the balance upside down. In Fig. 9.4, we now see the weight is in the middle, hanging down, and the supports are the edge of a table at one end, and a finger at the other. Comparing this with the first balance diagram, it is plausible that the table edge and the finger both support one half of the central weight, and a slightly more complicated experiment will confirm this also. If we now raise the weight, using our finger at the right hand end, it appears that a finger at twice the distance from the fulcrum (the edge of the table), but pulling upwards with half the 'weight', will lift the weight at the centre. Similarly, we can take the 3-unit–1-unit balance and invert it. The forces on the finger and the edge of the table are respectively 1 and 3 (Fig. 9.5) and we can test by experiment that lifting the finger with a force equivalent to 1 unit of weight lifts 4 units of weight four times closer to the fulcrum.

Exchange the finger and the fulcrum (Fig. 9.6), and ¾ of the effort is needed to lift a weight ¾ of the distance of the finger from the fulcrum.

These arguments are extremely convincing, not least because they fit experience very closely. Needless to say, it is usually helpful when

Fig. 9.6

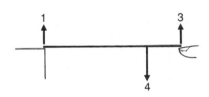

Problem 9C   *Another balancing act*

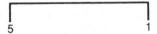

Fig. 9.7

Figure 9.7 shows two weights, of 5 and 1 units, at the ends of a light rod. How can the point at which the rod must be supported for the weights to balance be 'logically deduced' by following these simple rules?

1. The centre of gravity of two equal weights is at the point half way between them.

2. When deducing centres of gravity, a set of weights may be replaced by a single weight, at their centre of gravity, equal to their total weight.

3. A single weight may be replaced, for the purposes of deducing centres of gravity, by two or more weights which have the same centre of gravity and the same total weight.

solving a problem to know the solution you are supposed to reach
before you get there! Archimedes certainly often knew exactly what
he was after, because he used mechanical experiments to suggest
solutions, and methods of proving them.

Just how Archimedes achieved this was revealed by the discovery,
in 1906, in Constantinople, of a long lost treatise, the *Method* of
Archimedes. This is how Archimedes explained his procedure to his
friend Eratosthenes (most famous for accurately calculating the size
of the earth):

> I thought fit to write out for you . . . the peculiarity of a certain method, by
> which it will be possible for you to get a start to enable you to investigate
> some of the problems in mathematics by means of mechanics. This procedure
> is . . . no less useful even for the proof of theorems themselves; for certain
> things first became clear to me by a mechanical method, although they had to
> be demonstrated by geometry afterwards. But it is of course easier, when we
> have previously acquired by the method some knowledge of the questions, to
> supply the proof than it is to find it without any previous knowledge.

The same could be said by an increasing number of mathematicians
today who use that wonderful mechanism, the electronic computer,
to learn as much as they can about a problem before they attempt a
mathematical proof.

We can do the same as Archimedes, albeit at a lower level, and
discover 'facts' about levers and balances and other simple mechan-
isms before we try to prove anything. And the results of our experiments will indeed be convincing. It is difficult to imagine a world in
which balances and levers did not behave in the usual, almost
common-sensical, manner and it is easy to understand that these laws
were thought to possess an absolute truth, comparable to the truths
of elementary geometry.

Such simple 'laws' are also surprisingly powerful. Here is an
example. Suppose that someone is climbing a ladder. When is the
ladder most likely to slip? For the sake of simplicity, let us suppose
that a man carrying a heavy load is climbing a light aluminium
ladder, so that we can ignore the weight of the ladder. To make it
even simpler, suppose that the ladder is at the (hazardous) angle of
45° to the vertical and horizontal (Fig. 9.8). Call the length of the
ladder d, and let x be the distance of the man from the base. Think
of the ladder with the man on it, temporarily stationary, as one
object. There is a force, call it P, which pushes horizontally against
the ladder at its top end, and another force (of friction) pushing
upwards parallel to the wall at the top end which helps to stop in
sliding down; call it R.

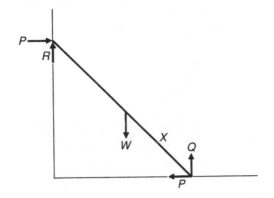

Fig. 9.8

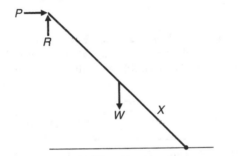

Fig. 9.9

At the bottom end there is a force parallel to the ground which pushes to the left. There are no other horizontal forces, so this force must equal P. There is also a force, say Q, upwards against the foot of the ladder. The two upwards forces must balance the weight of the man and his load, W, so $R + Q = W$.

That is one relationship between the various forces and the weight W. Now think of the ladder-plus-man-and-load as a lever which is able to rotate about its bottom end. It does not actually rotate at all, so the effect of the weight W in rotating it anticlockwise must be balanced by the effect of both forces R and P in rotating it clockwise (Fig. 9.9). These tendencies to rotate are proportional to the forces and the distance from the foot of the ladder to the one of the force. So the contributions of R and P are proportional to d and the contribution of W is proportional to x. Therefore

$$Rd + Pd = Wx \quad \text{and} \quad R + P = Wx/d.$$

Problem 9D   *The man as the fulcrum*

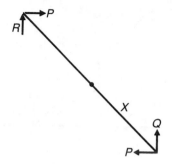

Fig. 9.10

Instead of treating the ladder as a lever, with the bottom end as the fixed point, we could treat it as a balance, with the man as the fulcrum (Fig. 9.10). How could we deduce the same result by this approach?

Problem 9E   *Find the centre of gravity*

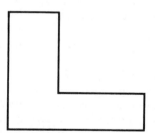

Fig. 9.11

Figure 9.11 shows a rectangle with a rectangular corner cut out; alternatively, it is two rectangles stuck together. It is possible to find the centre of gravity of this L shape with the use of a straight edge only – in other words, by drawing straight lines but with no measurement at all. How?

In other words, the sum of the two frictional forces, P acting horizontally at the foot of the ladder and R vertically at its top end, which
stop it sliding down the wall, are proportional to the distance of the
man-and-load from the bottom of the ladder.

The higher he climbs, the more likely the ladder is to slip. This fits
our experience. Standing on the bottom rung of the ladder you feel
no apprehension that the ladder will slip. Stand near the top, and you
do not feel so confident. The same ideas indicate the solution
to another problem: when you place clothes on a clothes-horse,
resting on a smooth linoleum floor, the clothes-horse is more likely
to slip if the clothes are hung over the top than if they rest on the
lower rungs.

Aristotle had supposed that all mechanics was ultimately based on
the properties of the lever and this was the extent of mechanics from
the time of the Greeks right up to the Middle Ages and beyond.
The weakness of this approach is that the balance and the lever
are about *parallel* forces. Some forces, or weights, pull down, some
pull up.

The next great step forward came with the solution of a puzzle
which had baffled Pappus of Alexandria, and all the Greek geometers,
and many other mathematicians up to and including Leonardo da
Vinci. In Fig. 9.12, a heavy object is being pulled up a slope. Everyday
experience tells us that, the shallower the slope, the less effort is
needed to haul the object; but how great is that effort? In contrast
to the case of the balance and lever, common sense does not tell
us. More subtle reasoning, which makes greater assumptions, is
required.

It was first solved by an anonymous pupil of Jordanus ($c.$ 1225).
Jordanus wrote several books on arithmetic and geometry, but nothing as original as his pupil's elegant argument, which was based on
considering *two* weights, each tending to slide down opposite sides of
a wedge, but prevented from doing so by a string joining them, which
slides over the top point of the wedge (Fig. 9.13).

Fig. 9.12   Fig. 9.13

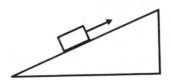

Problem 9F   *Where is the quadratic?*

In the middle ages, quadratics could be solved but not cubics or equations of higher degree. Naturally, quadratics occurred frequently in problems. Jordanus proposed the problem of separating a number into two parts, such that the sum of the squares of the two parts was another given number. For example, the first original number might be 12 and the sum of the squares of its parts, 80. What is the connection between this problem and quadratic equations?

Nowadays, we could make this idea the basis for an actual physical experiment, using two trolleys with smooth axles, each carrying a load, and with a very smoothly turning pulley at the top of the wedge. Such experiments, however, depend on modern technology to get rid of the effects of friction, and it is highly unlikely that Jordanus' pupil even considered an actual experiment. He proposed, rather, a *thought experiment*. Just as we have considered how levers and balances 'ought to behave' if they are to fit common sense, so early scientists tended to think first about how the world 'ought to be', judged by common experience. The purpose of a *thought experiment* was to allow the scientist to reason about the situation.

The very idea that you could actually set up experiments, and make measurements, supplementing common sense by extra information, before you started to speculate, had not occurred to them. This was unfortunate, because common sense told them very little, for example, about the behaviour of moving objects, so much of their speculation took place in a vacuum. To be fair to these early scientists, however, it has to be admitted that thought experiments have remained an important feature of scientists' thinking, and that modern scientists, too, when they imagine situations which they cannot actually handle by experiments, may argue and disagree with each other over the results of thought experiments, and come to contrary conclusions, much like their medieval forebears.

Anyway, the arguments based on the picture of the weights on the wedge were successful. We will look at the beautiful argument of the Dutchman, Simon Stevin (1548–1620). Stevin was responsible for popularizing the use of decimal fractions. He also invented the first sand yacht, a carriage with sails that raced a horse along the shore and won. More seriously, he was chief engineer and quartermaster general to the Prince of Orange, and wrote on the theory of fortifica-

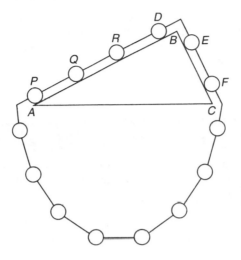

Fig. 9.14

tions, a popular Renaissance topic, and military engineering, so he naturally had an interest in the mechanical questions of the day.

Stevin imagined two slopes, forming a hill, as it were, one side of which was twice as long as the other (see Fig. 9.14), so $AB = 2BC$. On the slopes he imagined equally spaced identical spherical weights; the number was of course proportional to the lengths of AB and BC, so there were twice as many on AB as on BC. There is no friction between the spheres and the slope, so each sphere is supported solely by the tension in the string.

Finally – and this was a stroke of genius – Stevin added a loop of identical weights, hanging from A and C to complete a ring of spheres. At A, B and C are fixed smooth guides over which the cord can run freely without getting caught. He then started his argument by asserting that since perpetual motion was impossible (a fact that Leonardo and Cardano had asserted) the ring as a whole would not move. Therefore, subtracting the symmetrical loop of weights at the bottom, the weights on either slope effectively balanced each other.

He concluded that the weights E and F balanced the weights P, Q, R and D. Figure 9.14 is based on Stevin's original diagram, which he reproduced on the front of his book *Hypomnemata Mathematica* with the motto, 'Wonder en is gheen wonder', 'The magic is not magical', indicating that what seemed to be magical had now been reduced to reason. Stevin immediately deduced from this argument the weight hanging vertically which would balance a given weight on a slope (Fig. 9.15). By his argument, $E/D = BC/AB$, or – as we would say, using trigonometrical ratios – the weight is $D \sin \theta$.

Stevin next tackled the problem of the tensions in two strings which support a weight, and which are perpendicular to each other (Fig. 9.16). Stevin could have deduced this also directly from his previous demonstration (in fact he used a rather different argument). Returning to Fig. 9.13, consider one side of it, that is, the situation of just one sphere, weight W, supported by the pressure against the slope, which will be perpendicular to the slope since there is no friction, and the tension in the string (Fig. 9.17). The latter is equivalent, as we have seen, to a weight of $W \sin \theta$, and the pressure against the slope is unknown.

However, the situation of the single sphere would be identical if we replaced the tension in the string by a solid sloping wall, against which the sphere rested, and replaced the pressure from the original

Fig. 9.15

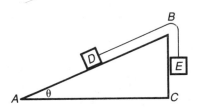

Fig. 9.16

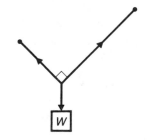

Fig. 9.17

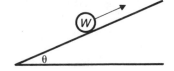

slope by a new string (Fig. 9.18). By Stevin's previous argument, the tension in the new string will be $W \cos \theta$, and the pressure against the new slope, since the position of the sphere is unchanged, will be $W \sin \theta$ as before. We can therefore forget both pressures, and consider the particle as supported by two strings, with tensions $W \sin \theta$ and $W \cos \theta$, as in Fig. 9.19. This is the same as saying that the force resulting from the two tensions in the strings is represented by the vertical diagonal of the parallelogram in Fig. 9.20, in which the lengths of the lines are now proportional to the strengths of the forces. In general, two forces represented in strength and direction by two sides of a parallelogram are equivalent to a single resultant force, represented by the diagonal of the parallelogram.

Fig. 9.18

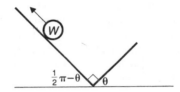

Fig. 9.19

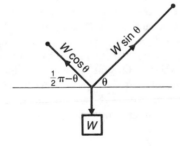

Fig. 9.20

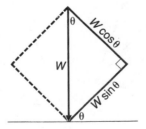

Problem 9G   *Moments and the parallelogram of forces*

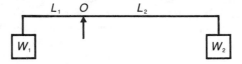

Fig. 9.21

The idea of *moments* is based on the ideas of balances and levers. If the two weights on the balance shown in Fig. 9.21 are in equilibrium, then $L_1W_1 = L_2W_2$. L_1W_1 is called the *moment* of the weight W_1 about the point of balance, O. In general, we can calculate the *moment* of any force about any point: it is just the strength of the force, multiplied by the perpendicular distance of the point from the line along which the force acts. In Fig. 9.22, force F is represented in strength and direction by the arrow AB. Its moment about the point P is equal to $AB \times PD$, which happens to be (geometrically speaking) double the area of the triangle ABP.

If two given forces are equivalent to a third force – their *resultant* – then the sum of the moments of the two forces about any point, P, will be equal to the moment of the third (resultant) force about P. We can use this idea to check the parallelogram law of forces (or, by working backwards, actually deduce it). In Fig. 9.23, forces represented by OA and OB are equivalent to the resultant force OC. Therefore, the total moments of OA and OB about any point we care to choose should equal the moment of OC about that point.

Let us choose, for simplicity, the point B itself. The moment of OB about B is zero, because it goes through B and so its distance from B is zero. What are the moments of OA and OC about B, and why are they equal?

Fig. 9.22   Fig. 9.23

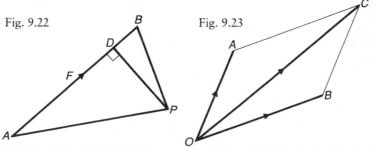

*

The next great step forward, or rather, several giant steps, were taken by Galileo Galilei (1564–1643). Prior to Galileo, in 1543, Tartaglia had published a translation of two of the mechanical treatises of Archimedes, making them widely available to Renaissance engineers and natural philosophers. These translations had the profound effect of making mechanics and engineering a respectable subject of study by professors and scholars. The ancient Greeks themselves had regarded engineering as a craft and, as such, demeaningly practical. True knowledge, taught Plato, came from contemplation and the exercise of the intellect. It was this message which caused subsequent philosophers to emphasize thought experiments over real experiments, with such mixed results.

Archimedes, however, had brilliantly *combined* his own thought experiments with actual mechanical experiments and purely mathematical reasoning, with stunning results, which Tartaglia did not fail to point out: in Venice, only a few years earlier, a new ship of the Venetian navy had sailed out into the lagoon, fully armed, and at once capsized and sank. The works of Archimedes explained why – its centre of gravity was too high. Archimedes had also founded the science of hydrostatics, and explained how to calculate the centres of gravity of complicated floating bodies.

In 1577, Guidobaldo del Monte published a treatise on mechanics which discussed mathematically the five simple machines: the lever (balance), the pulley, the axle (gear), the wedge and the screw, on Archimedean principles. Galileo was a pupil of Guidobaldo.

Galileo solved the problem of the weight on the slope, by appealing to the properties of levers. He supposed that a weight at F in Fig. 9.24 is held on the slope GH by the arm of a lever whose fulcrum is at B,

Fig. 9.24

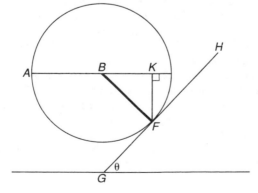

the centre of a circle touching the slope at *F*. The force needed to stop the weight rolling down the slope is equal to the force needed to stop the lever turning clockwise. It was understood that the force exerted by the weight at *F* on the lever *BF* was the same as that which would have been exerted by an identical weight hanging from *K* by a thread, and that this was proportional to the magnitude of the weight at *F* and the distance *BK*: in modern terminology, the product of weight and the distance *BK* is the *moment* of the weight about *B*. The ratio *BK/BF* is, as it happens, cos $\widehat{KBF}$, which is the sine of the angle between the direction of the slope *GF* and the horizontal – the same result that Stevin reached.

Galileo's approach to the weight-on-the-slope problem makes clear that it is, contrary to first appearances, intimately related to the old and familiar problems of levers and balances. Almost as closely related are the general problems of the resultant of two forces, solved by the parallelogram of forces. Consequently, we could say (with only a little poetic licence) that, when seen in the right perspective, these problems are also soluble by common sense. They certainly follow logically from the simplest assumptions and a few thought experiments.

The next giant step, taken by Galileo, is a different matter altogether. Mathematicians had long been interested in matters military. Archimedes had built military machines, and of course many mathematicians are employed by the military today. The infant science of ballistics, or the flight of projectiles, was especially important. However, there were severe difficulties in the early investigations. In the first place, it was by no means easy to see exactly what happened to a projectile flying through empty air. We may find it obvious nowadays because we have seen time-delay photographs, or observed the flight of a golf ball on television, or simply read popular science magazines. Renaissance scientists had to rely on a combination of naked eyesight, and their own philosophical preconceptions which more often than not distorted the observations of their eyes – as any modern psychologist would have predicted – so that they saw what they expected to see, rather than what was actually there.

Tartaglia in an early work illustrated the then popular view of how a projectile moved through the air. It first moves in a straight line, then in a curved portion, which turned it back towards the earth, and then finally again in a straight line. No wonder mathematicians had difficulty in making sense of such paths!

Tartaglia himself argued that, on the contrary, the path of a cannonball started to bend towards the earth from the moment it left the gun barrel, but the gunners were not impressed. In fact, the Duke of Urbino had made some trials in which a culverin, aimed horizontally at a target 200 paces away, did hit the target. Of course, we know, and Tartaglia argued, that the vertical deviation in this test would be too small to be spotted by eyesight. This episode illustrates the essential fact – essential for an understanding of the slow growth of actual experiments in the development of science – that effective experiments so often depend on successful technology: if the technology is primitive, so will be the experiments.

Galileo made his breakthrough in several stages. Firstly, he decided that the path of a projectile on a smooth steeply sloping plane was effectively the same as that of a projectile flying through the air. He, as it were, laid the path slightly on its side. Given allowances for the friction of the sloping board, Galileo was brilliantly right.

Now that he could see the path laid out in front of him, he used the result of some experiments he had done with smooth spheres rolling down a strip of wood in which, he described, '. . . a channel, a little wider than one finger, was hollowed out. It was made quite straight and, in order that it should be polished and quite smooth, the inside was covered with a sheet of parchment as glazed as possible. A short ball of bronze that was very hard, quite round and well-polished, was allowed to move down the channel.' Galileo timed the journey of the bronze ball down the channel, against his own pulses. Another example of primitive technology: as it happened, it was Galileo who invented the pendulum clock by observing the swinging of a church lamp on a long chain, and a very short pendulum could have been used to time his rolling balls.

Problem 9H   *Is it time or distance?*

Medieval philosophers had disagreed on the question of the speed of a falling object. Some had claimed that it was proportional to the time elapsed, others that it was proportional to the distance travelled.

As late as 1604, Galileo himself still believed in the wrong answer to this question. But, according to Galileo's experiments with the ball rolling down a chute, the distance travelled is proportional to the square of the time, or

$$d = pt^2,$$

where p is some constant. How does this settle the medieval disagreement?

Having repeated that experiment many times, 'we made the ball fall through only a quarter of the length of the channel, and found that the measured duration of fall was always equal to half of the other.'

In other words, halve the time, and the distance is quartered; double the time, and the distance travelled is multiplied by 4 (treble the time and it is multiplied by 9). The distance travelled was proportional to the square of the time taken.

Galileo now made another brilliant move. He supposed, correctly, that the motion of a projectile consisted of two entirely independent parts. On the one hand, it had a tendency to fall backwards to the centre of the earth, according to the law he had just discovered. This was its vertical motion. On the other hand, it had an independent horizontal motion, which was constant. Performing these two motions together, the path of the projectile, shown in Fig. 9.25, thrown off a cliff at B, is one half of a parabola. In the dialogue in which Galileo introduced his arguments (in the mouths of Salviati, Sagredo and Simplicio) Salviati acknowledges that he has ignored the resistance of the air, and the fact that the earth is a sphere and so 'vertical' lines are not parallel but converge to the centre of the earth. Galileo knows that he is making an approximation, but a very close one.

Galileo also realized that the speed with which a sphere rolls down a plane slope depends on the angle of the slope, because that determines the force on the sphere, parallel to the direction of the slope. Galileo explained this, through the mouth of Salviati, by describing the following experiment, which perfectly illustrates his remarkable imagination and insight. (See Fig. 9.26.)

Fig. 9.25

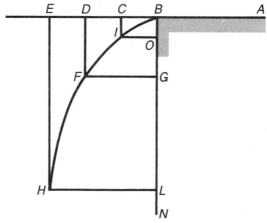

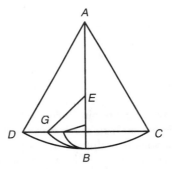

Fig. 9.26

A small lead weight at *B* is attached to the pin at *A* by a thread *AB*. Pulled to one side at *C*, and released, it swings in an arc to *D*, apart from 'a small deficiency . . . precisely due to the resistance of the air and of the thread'. Therefore the *impetus* which the ball acquires in falling from *C* to *B* will be sufficient to carry it up to *D*, with the given qualification. Next, Galileo fixes a nail at *E*, so that the thread, as the ball swings, is caught against *E* forcing the ball to rise in the arc *BG*. Galileo observes, 'Now, gentlemen, you will see with pleasure that the ball attains the horizontal at the point *G*. The same thing would happen if the nail were fixed lower . . . The ball will . . . always finish its ascent on the line *CD*,' unless the nail were too low for the ball to reach *CD*, in which case, Galileo observes, the ball and thread wrap themselves round the nail. Galileo then deduces that the *impetus* acquired in *descending* arcs such as *DB* or *GB* is in every case equal.

Galileo went on to conclude that, when balls roll down slopes of different steepness from the same height, although the ball on the shallowest slope takes longest, they all acquire the same speed at the same horizontal level.

A wide chasm separates the laws of the lever and the balance, or even the sphere on a slope, from the law of motion of a projectile. Galileo's analysis is at once highly imaginative and highly practical, based on experiments whose results could not be predicted by common sense. There is something mysterious about the law of the parabola, which typifies all subsequent laws of physics. The laws are often very simple – suspiciously simple. Pictures of them always look more like Galileo's parabola than the weird curve illustrated by Tartaglia. Even when they are much more complex – and modern physical theories are very complex indeed – they are still simple *relative to* the subtle phenomena that they describe. Although they are approximations, as Galileo realized from the start, they are very good approximations indeed. It does seem that, by such simple mathematical equations and formulae, we are getting very close to some simple ultimate truths about the universe: but why?

Einstein once said that the most incomprehensible thing about the universe is that it is comprehensible. Eugene Wigner, a Nobel-prize-winning physicist, wrote a famous paper, titled 'The Unreasonable Effectiveness of Mathematics in the Natural Sciences', in which he asserted that 'the enormous usefulness of mathematics in the natural sciences is something bordering on the mysterious and . . . there is no rational explanation for it,' and he concluded that 'it is difficult to avoid the impression that a miracle confronts us.' Wigner no doubt had in mind advances in physics in this century, such as relativity theory and especially quantum theory, which have been extraordinarily successful, and yet are also quite incomplete, as well as being far from the intuitions of common sense. If physicists really are steadily approaching some ultimate truth about nature, then it can only be said that, as they see it more and more clearly, it becomes more and more bizarre.

Science, however, does not comprise only such weird theories, nor is mathematics only used to describe the world precisely and accurately. Far from it, much mathematics today is used to construct models of the world which are not so very accurate, do not necessarily reveal any profound truth, and yet are highly effective, as we shall see next.

Problem 91 *Galileo and the falling weights*

Aristotle had taken the common-sense view that a heavier weight fell faster than a lighter weight – a plausible conclusion if you are comparing the fall of a feather and a lump of lead. Subsequently, Aristotle's view was taken as gospel by generations of natural philosophers. Galileo is supposed to have dropped different weights from the top of the Leaning Tower of Pisa, in order to prove that Aristotle was wrong. Whether or not that story is true, Galileo did propose the following thought experiment, as an argument to undermine Aristotle's conclusion. Imagine, said Galileo, that a 10-unit weight and a 1-unit weight are joined by a long loose string. They are dropped together from a great height . . . Query: what was the rest of Galileo's argument?

Solutions to problems, Chapter 9

9A: *Problems from Diophantos*

(These are Problems 24, 25 and 27 in *The Penguin Book of Curious and Interesting Puzzles*.)

 1. The first number is 98, the second, 94.

 2. The sum of the four totals will include each number three times: therefore, the sum of the four numbers is $\frac{1}{3}(22 + 24 + 27 + 20) =$ 31. The required numbers are therefore 9, 7, 4 and 11.

 3. This is Diophantos' solution, which typically finds a particular solution by making simplifying assumptions.

Take a square and subtract part of it for the third number; let $x^2 + 6x + 9$ be one of the sums, and 9 the third number. Therefore the product of the first and second numbers is $x^2 + 6x$; let the first be x and the second $x + 6$. By the two remaining conditions, $10x + 54$ and $10x + 6$ are both squares. Therefore we have to find two squares differing by 48 . . . The squares 16 and 64 satisfy this condition. Equating these squares to the respective expressions, we obtain $x = 1$, and the numbers are 1, 7 and 9.

We might now notice that Diophantos' solution can be written in the form 1, $2n^2 - 1$ and $2n^2 + 1$, where $n = 2$. A very little algebra shows that this solution works for any value of n.

9B: *Balancing a triangle*

Because the centre of gravity of the right-hand triangle is one third of the way from the midpoint of CD towards B. It will therefore be directly under the point on the horizontal arm which is one third of the way from O to B. Since A and B are equal distances from O, a weight one third of that of the triangle, hanging from A, will balance the triangle.

9C: *Another balancing act*

The starting position is represented like this. The space between the 5 and 1 weights has been divided into $5 + 1 = 6$ parts:

● ● ● ● 5 ● ● ● ● 1

By Rule 3, replacing the 5:

1 ● 1 ● 1 ● 1 ● 1 ● 1

We could now simply say, 'By symmetry, the centre of gravity is at the centre', but following the rules strictly, we continue to apply them like this:

By Rules 1 and 2:

$$= 1 + 2 \qquad\qquad = 1 + 2$$

By Rule 1 and 2 again:

● ● ● ● ● 6 ● ● ● ● ●

As expected, the two weights are equivalent to a total of 6 units at the point dividing the line joining them in the ratio $1:5$.

9D: *The man as the fulcrum*

Considering the ladder as a balance, with the man at the centre, we can see that the effects of the forces P and R at the top are proportional to $R(d - x)$ and $P(d - x)$. To these must be added the effect of the force P at the bottom, proportional to Px. These are opposed by the effect of force Q, proportional to Qx. Therefore,

$$R(d - x) + P(d - x) + Px = Qx,$$

and so $Rd + Pd = Qx + Rx = Wx$ (since $R + Q = W$), the same result as before.

9E: *Find the centre of gravity*

Yes, it is, by thinking of it as split into two rectangles, in two different ways (Fig. 9.27). In the first figure, the centres of the two rectangles are found by simply drawing their diagonals, and the centre of gravity of the whole figure lies somewhere on the line joining them.

Fig. 9.27

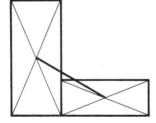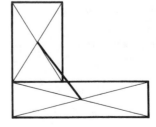

Similarly, in the second figure, the centre of gravity of the whole figure lies on the different line joining two centres of gravity. When the two figures are combined, the centre of gravity is where the two lines cross. Note that the centre of gravity could lie outside the boundary of the shape.

9F: *Where is the quadratic?*

Call the parts x and y. Then,

$$x + y = 12 \quad \text{and} \quad x^2 + y^2 = 80.$$

Therefore $x^2 + (12 - x)^2 = 80$. Simplifying, we get the quadratic equation

$$x^2 - 12x + 32 = 0.$$

Alternatively, we could argue that knowing the sum of two numbers and their product is equivalent to solving a quadratic; here we know the value of $x + y$, and we can find xy because it is

$$\tfrac{1}{2}[(x + y)^2 - (x^2 + y^2)] = \tfrac{1}{2}(144 - 80) = 32.$$

9G: *Moments and the parallelogram of forces*

The moment of the force OA about B is the length OA multiplied by the perpendicular distance BA', which is double the area of the triangle BOA, or the area of the parallelogram $BOAC$ (Fig. 9.28). The moment of the force OC about B is the length OC multiplied by the perpendicular distance BC', which is also equal to the area of the parallelogram. Hence the moments are equal and, since one of them is clockwise and the other anticlockwise, they balance.

Fig. 9.28

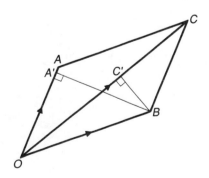

9H: *Is it time or distance?*

If the falling object has travelled $d = pt^2$ after time t, and $D = pT^2$ after a later time T, then its average speed will be

$$(D - d)/(T - t) = (pT^2 - pt^2)/(T - t) = p(T + t).$$

If the times t and T are very close together, then $p(T + t)$ will be very close to $2pt$, and the speed 'at an instant' will be proportional to the time elapsed, and not to the distance travelled.

9I: *Galileo and the falling weights*

If Aristotle was correct, then the heavier weight would fall faster, and so the string would tauten, and soon the heavier weight would be pulling the lighter weight, which naturally travels more slowly, after it. This suggests that the attachment of the lighter weight would slow down the heavier weight, and yet the two weights together, weighing 11 units, ought surely to travel *faster* than either of the weights separately, according to Aristotle.

Chapter 10
Mathematics in science: approximate models

In Chapter 7 we saw that we could estimate the number of animals in an area by capturing a number of animals, marking and releasing them, then capturing some animals again, and counting the number of animals that had been *recaptured*. The estimate for the total population is given a special name, the *Lincoln index*. Why does it have a name at all? Why is it so special? The Lincoln-index method is only so special and so simple because it makes so many assumptions: that the marked and unmarked animals are randomly spread through the region, and equally likely to be captured; that no animals are born, or die, between the original capture and the recapture, and that no animals leave the region or enter it; and that the proportion of marked animals in the recaptured sample really is identical to the proportion of marked individuals in the whole population.

Put like that, it is obvious that the Lincoln index itself is no more than an approximation, matching the simplified and approximate picture on which it is based. Here is another example of the same phenomenon: suppose that you are a forest ranger and that your tasks include estimating the amount of timber in a given area. How do you do this? By exhaustive measurement of every tree in the area? Impossibly time-consuming. By measuring a sample of trees chosen at random? This is much more plausible, but how will you randomly select the trees?

If the trees are quite widely spaced, then a simpler method is this: select several points at random in the forest. Viewing the forest from each point in turn, count the number of visible tree trunks that subtend at angle greater than some suitably chosen angle, a. This count can be done efficiently by constructing an instrument for the purpose. If the average number of trunks counted is N, then the mean area of trunks in a unit area of forest is proportional to $N \sin^2 a$. In practice, if a is chosen to be $1° 44'$, then $10N$ is an estimate of the total area of the cross-sections of the trunks, measured in square feet, per acre. In Fig. 10.1, nine tree trunks are shown, of which three are only partly visible from the chosen point. Of the remaining six, just two subtend angles greater than the chosen angle of $15°$.

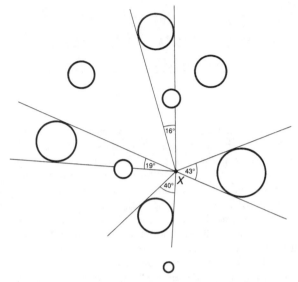

Fig. 10.1   *X* is the chosen point. The chosen angle is 15°.

What simplifying assumptions have been made in this example? The mathematical calculation assumes that a tree has a circular cross-section, and that a vertical view of the forest will show the trees as perfect circles of varying size whose centres are randomly distributed. The first assumption is quite a good one, the second is more suspect. The location of trees depends on many factors, and is only random to a first approximation.

Both the Lincoln index and the formula for counting trees are examples of *models*. A model 'mimics relevant features of the situation being studied'. Features that are not relevant are simply omitted. That means that a road map, which is a model of a part of the earth's surface, usually omits geological features, just as geological maps usually omit roads, and maps of climate omit both.

Among the features which are initially judged irrelevant there may well be features which the scientist would like to include, but cannot, at least in a first, very simple model, because they would make it too complicated to handle. However, after an initial model has been constructed and tested, then it may well be possible to add in features which were, despite their obvious importance, reluctantly omitted at the start.

Problem 10A   *Olbers' paradox*

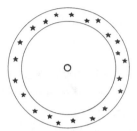

Fig. 10.2

It sometimes happens that a very simple and apparently plausible model is quite obviously wrong. The model then has no practical use, but it does force scientists to think hard, in order to discover what is wrong with it. Olbers (1758–1840) considered the amount of light that we would receive on earth from the stars, assuming them to be distributed evenly (on a large scale) throughout the universe, and infinite in number. On the one hand, he argued, the number of stars in a thin shell between two spheres centred on the earth (Fig. 10.2) would be proportional to the square of the radius of the spheres. On the other hand, the perceived brightness of those stars would be *inversely* proportional to the square of their distance from us. To what paradoxical conclusion did this argument lead him?

The simplest arguments about levers assume that the levers themselves weigh nothing. Likewise the simplest arguments about balances assume that the arms of the balances are weightless. Although Stevin and Galileo would not have recognized the term 'model', they were in fact both considering simplified models. Stevin, in his thought experiments, imagined that there was no friction between the small spheres and the surface of the slope, and that the thread at the top passed over the fixed point completely smoothly. He knew, as we know, that this will not be so in a practical model. Indeed, the question of perpetual motion would not arise in a practical model, and his whole argument would not start.

Likewise, Galileo tried to make the channel down which the brass sphere rolled as smooth as possible; but he knew that it was not perfectly smooth, just as he must have been aware that his own pulses against which he timed its descent, were not perfectly even. In

Problem 10B   *The flying fish*

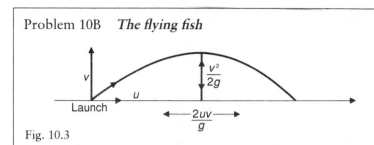

Fig. 10.3

A simple model can also often be used to show that a phenomenon is not as simple as supposed. Consider a projectile, launched from a point on a horizontal plane, whose horizontal velocity at launch is u and whose vertical velocity is v. Let g be the acceleration due to gravity. Then the range – the horizontal distance from the point of launch to the point where it strikes the ground – is $2uv/g$; and the maximum height reached is $v^2/2g$ (see Fig. 10.3).

Now suppose that a flying fish is observed to leap from the smooth surface of the ocean at an angle of 20° to the horizontal, reaching a maximum height of 0.5 metres and landing 7 metres from its launch point. How can you infer that it is indeed 'flying', in some manner, rather than behaving merely as a projectile?

calculating the effect of a man carrying a load up a ladder, we explicitly made quite a list of assumptions, including the implausible idea that the ladder itself had zero weight.

Mathematical models are continually invoking ideas of infinitely smooth surfaces, weightless strings, weightless beams, perfectly spherical balls, projectiles flying through airless space, gases which are perfectly compressible and liquids which are perfectly incompressible, and so on. The purpose of such simplifications is, in theory, to understand the world better *despite* the oversimplification, which you hope either will not matter or will be corrected when you construct a second (better) model.

Francis Galton (1822–1911) started off as an explorer, who then turned his interest to meteorology and then anthropology. He had observed and wished to explain scientifically the enormous differences in abilities between individuals and groups. His most famous researches appeared in *Hereditary Genius* in which he argued, by analysing the family histories of many talented individuals, that their abilities were largely inherited. This research led him to speculate on the fate of such families over many decades:

> The decay of the families of men who occupied conspicuous positions in past times has been a subject of frequent remark, and has given rise to various conjectures. It is not only the families of men of genius or those of the aristocracy who tend to perish . . . The instances are very numerous in which surnames that were once common have since become scarce or have wholly disappeared . . .

He continued, revealing the theme that made him a founding father of the eugenics movement, and makes this a problem of historical importance (the assumptions he makes about the inferiority of the 'proletariat' were typical of his time):

> the conclusion has been hastily drawn that a rise in physical comfort and intellectual capacity is necessarily accompanied by diminution in 'fertility' – using that phrase in its widest sense and reckoning abstinence from marriage as sterility. If that conclusion be true, our population is chiefly maintained through the 'proletariat', and thus a large element of degradation is inseparably connected with those other elements which tend to ameliorate the race.

Galton wished to know how likely it was that a family name would eventually die out, given suitable assumptions about the number of children produced in each generation. He was not at first able to solve the problem, so he proposed it as a question in *The Educational Times*, which was famous for its mathematical problems and solutions, often contributed by the most eminent mathematicians of the day.

Only one – completely incorrect – solution being received, he proposed it to the Reverend Henry William Watson, a mathematician and early mountaineer. Watson did solve the problem, apart from a correctable slip, by making 'suitable' simplifying decisions. He was interested in the persistence of male lines, so he considered only sons. He assumed that a typical male would have no sons with probability p_0, one son with probability p_1, two sons with probability p_2, three sons with probability p_3, and so on. He further assumed that all males were typical, so the same probabilities applied to every male in the population – a very simplified assumption of course, since, as Galton implied in his own comments, males in different classes did have, on average, different numbers of sons.

However, these simplifying assumptions did make the problem solvable, by the method of generating series, which goes back to Euler. Watson showed that the probability of the male line ultimately becoming extinct is the smaller real positive solution in x of the equation

$$x = p_0 + p_1 x + p_2 x^2 + p_3 x^3 + \cdots.$$

Clearly this series soon stops: even in Victorian England, no family had more than a dozen or so sons.

In 1931, A. J. Lotka calculated the probability that a male line in the United States of America would die out. He used statistical data to determine that $p_0 = 0.4982$, $p_1 = 0.2103$, and so on, leading to this equation:

$$x = 0.4982 + 0.2103x + 0.1270x^2 + 0.0730x^3$$
$$+ 0.0418x^4 + 0.0241x^5 + 0.0132x^6 + 0.0069x^7$$
$$+ 0.0035x^8 + 0.0015x^9 + 0.0005x^{10}.$$

The smaller positive root of this equation is approximately 0.88, so the probability that a male line will die out is rather high, *if all the assumptions made along the way are sound.*

Just because a model simplifies the actual situation, it often happens that the same model will fit a variety of situations. The Galton–Watson model of generations is no exception. Imagine that you are a geographer who is studying the patterns of streams and their tributaries. Some streams have many tributaries, and working *backwards* to the source of each reveals a complex tree-like pattern (Fig. 10.4). Other streams form simpler patterns. How likely is it that as you work backwards up a stream, you will find that it branches? Or that its path leads unbranching to its unique source? One way to investigate

---

Problem 10C   *A dice simulation*

It is not difficult to simulate the generations of a family by throwing dice. Here is a very oversimplified model. You start by throwing a die. If you throw 1, 2 or 3, you do not get another throw and the game stops at once. This is equivalent to the very first generation having no sons. If you throw a 4 or 5 then you have another throw. This is equivalent to the first generation having 1 son. If you throw a 6, then you get two sons, and these rules apply to each of them.

What would you intuitively expect the probability to be that the game does *not* go on for ever? Using the values $p_0 = 3/6 = \frac{1}{2}$; $p_1 = 2/6 = \frac{1}{3}$; $p_2 = \frac{1}{6}$, and all other probabilities are zero: what is the unique positive solution of Watson's equation, in this case?

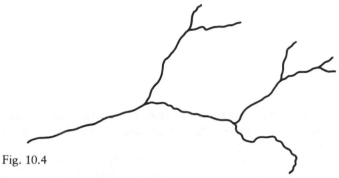

Fig. 10.4

the processes that determine stream patterns is to *imagine* that streams are actually formed by the same process as Galton's generations of males. Each stream has a probability that it will acquire tributaries and a probability that it will not – which makes this a simplified version of Galton's model.

The basic conclusion from this model is that if the chance of bifurcation is less than 50:50 (that is, if the average number of sons is less than 1) then the network must come to an end, sooner or later. But if the chance of bifurcation is greater than 50:50, then a proportion of networks will continue spreading for ever.

Problem 10D *How many sources?*

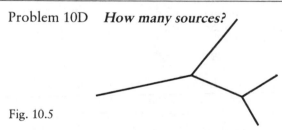

Fig. 10.5

Suppose that you are tracing a river back to its source, or sources. As an initial 'guesstimate', you suppose that the probability that the river will branch into two tributaries somewhere higher up is 50:50, or $\frac{1}{2}$, and that when you follow either of those tributaries (if there indeed are any) then they will also have a probability $\frac{1}{2}$ of branching again, and so on. The chance that there will be no bifurcation is $\frac{1}{2}$. The chance that there will be a bifurcation, and two tributaries, is also $\frac{1}{2}$ and on average one of these tributaries will not branch again but the other one will . . .

What is the probability that a river will have exactly three sources?

*

The capture–recapture problem, and the Lincoln index, is easy to handle because so many simplifying assumptions were made. The situation becomes much more complex when a population of, say, small animals is being actively preyed on.

The predator will flourish if there is a large population to be hunted, but if the predator is too successful, then the prey will be so reduced that the predator itself will suffer. The result is a complex interaction, in which the hunter has an interest in the survival of the hunted, as human beings have discovered in many parts of the world, where they prey on natural resources, which do not regenerate fast enough, and eventually fail completely, leaving the human population wishing that they had been less greedy.

This analogy with human behaviour suggests that, as so often, the same model can be used to describe many different situations. This is true: for example, a number of salesmen competing to sell their goods to a limited number of buyers. If too many salesmen successfully approach too few buyers, then the buyers may all be satisfied, and the salesmen suddenly find they have no customers and they go out of business. Ideally, the salesman serves a large number of customers, and is successful enough to make a living today while always leaving enough unsatisfied customers to give him work tomorrow.

The assumptions about the prey and the predator can be put into these Lotka–Volterra differential equations, in which x is the number of prey and y is the number of predators:

$$\frac{\mathrm{d}x}{\mathrm{d}t} = ax - cxy, \qquad \frac{\mathrm{d}y}{\mathrm{d}t} = fxy - ey.$$

Here, $\mathrm{d}x/\mathrm{d}t$ is the rate at which x changes with time. The term ax on the right-hand side represents the assumption that the growth of the prey population is proportional to its size: the larger it is, the faster it grows. The term $-cxy$ represents the assumption that the number of prey lost to predators will be proportional to the size of the prey population, and to the number of predators. Similarly, $\mathrm{d}y/\mathrm{d}t$ is the rate at which y changes with time. The assumptions represented here are that the predator population has a death rate, represented by $-ey$ and a growth rate which is proportional to prey and predator populations, fxy.

The rate at which the prey population changes, compared to the predator population, is the ratio of these two expressions. In other words,

$$\frac{\mathrm{d}y}{\mathrm{d}x} = \frac{fxy - ey}{ax - cxy}.$$

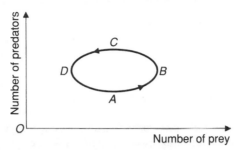

Fig. 10.6

This differential equation can be solved by using calculus. The result is shown in Fig. 10.6. Suppose we start at A at the bottom of this closed curve. The predator population is a minimum here, so the prey are flourishing. Therefore the prey increase as we move to the right, providing more food for the predators which increase also. At B, the prey has reached its maximum, but the predator population is still increasing, which reduces the prey population. At C, the predators are maximum in number and, continuing towards D, the number of both prey and predators falls. The prey are minimum in number at D, and thereafter start to multiply, finally reaching A, whereupon the cycle starts to repeat.

This is a very simple model, which has ignored some important factors. For example, we have assumed that the prey have as much food as they require, which is not plausible. As the prey increase in number, they will compete with each other for food, and this competition will tend to limit their number, even without the effects of the predators. To take this into account, we add an extra term to the first equation, which is now

$$\frac{dx}{dt} = ax - bx^2 - cxy.$$

The solution now looks very different. Instead of both populations oscillating indefinitely, the numbers of each tend to a limit (see Fig. 10.7). If the original numbers in each population are represented by

Fig. 10.7

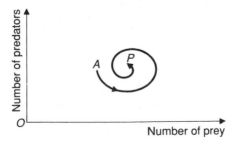

the point A, then the 'path' of the two populations is a spiral, which tends to a fixed point at P. As the two populations move towards P, they will oscillate in size, but the oscillations will get smaller and smaller, until they settle down in a kind of harmony. The predators are in no danger of wiping out their prey, but they themselves are limited in number; the number of prey is also stable.

What might we guess, from this conclusion, about the accuracy of these two models? Would you be really confident that, if food for the prey was very plentiful, the two populations would oscillate as in the illustration, perfectly repeating themselves? Would you expect the fixed point P to be immovably fixed, quite independently of changes in the weather, the behaviour of other species (which might be sharing their food stocks, for example), or the activities of human beings? Common sense suggests that these approximate models cannot tell us everything. Yet they may still tell more than we would have suspected without their aid, as well as providing the basis for less simplified second-generation or third-generation models.

To illustrate this point, Fig. 10.8 shows the result of a model of the interaction between a population of grazing animals (which could be the prey population in the last model) and the plants they eat. Suppose that the animals are deer which have just been introduced, in very small numbers, to the forest. By making suitable assumptions about the numbers of plants, the number of animals, the rate at which plants grow, the efficiency with which the animals graze the plants, and other factors, this model predicts that, over a period of years, the number of animals and the number of plants will oscillate

---

Problem 10E   *Where is the limit point?*

At the limit point in the second model (where the prey's food is restricted), neither x nor y will be changing. Therefore dx/dt and dy/dt will both be zero, and we have the equations below, from which the position of the limit point can be calculated:

$$ax - bx^2 - cxy = 0, \qquad fxy - ey = 0.$$

A straightforward calculation shows that, if $af > be$, then the limit point is given by

$$(x, y) = (e/f, a/c - be/cf).$$

If $af \leqslant be$, what is the limit point, and how would you interpret it?

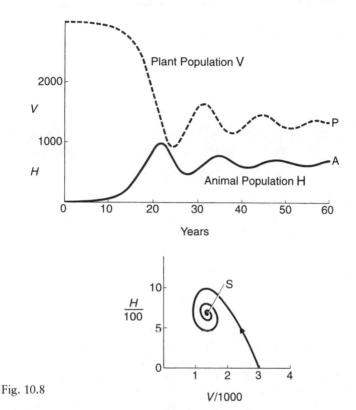

Fig. 10.8

towards a steady state, in which their ratio is constant. This is represented by the ratio of P to A in the first diagram, and the point S in the second.

Over the first few years, however, the situation is far from stable. The deer population starts by shooting up, while the plant population falls dramatically. This is what wildlife managers call a 'deer eruption'. The reaction of a novice manager (who had not experienced a deer eruption before) might be that, if the deer were not culled, there would soon be no plants left to support them, and therefore no deer either. Indeed, just such alarmist conclusions have been drawn by observers of elephant and lion populations in parts of Africa. The model strongly suggests that this fear is groundless. The deer and their food will stabilize in time, and both flourish together.

Stability is a typically important mathematical and practical concept, because it implies the possibility of planning. Economics is an

---

Problem 10F   *Stabilizing pigs*

If you were in charge of the modern equivalent of a Pig Marketing Board, how would you act to stabilize the price of pig-meat?

---

even more complicated discipline than ecology, because human beings are so unpredictable, and economic planning is notoriously difficult. Here is an example of deliberate economic planning to make a situation more stable than it would otherwise be.

Up to the time of the formation of the Pig Marketing Board in 1933 the fluctuations in the price of pork were proverbial, and the price varied in inverse ratio with the pig population. When the pig population was high the price of pig-meat fell, and when the fall had reached the point at which it was no longer profitable to produce them farmers hurriedly gave up their pigs and reduced their breeding stock. Ultimately the time arrived when pigs became scarce and dear once more with the result that the high profits made by those who still had pigs to sell were sufficient to attract farmers to begin breeding pigs again, and the whole cycle recommenced.

All the models that we have considered in this chapter have been rough and ready. They have all been obviously crude approximations, and no one supposes that they are anything more. This does not mean that they are useless – far from it – but it does mean that the answers they give to practical questions are also approximations. There is a pragmatic payoff here between the use of simple models which give good-enough answers which are good value for money, and the use of much more sophisticated models which are more powerful, but also more complex to use, perhaps requiring more advanced mathematics and the use of computers.

How do such powerful and complex models compare to the scientific theories of, say, physicists and chemists? The closest agreements between experimental measurement and theoretical prediction occur in the most advanced theories. In quantum theory, which is by far the most demonstrably accurate physical theory ever created, the differences may be only a few parts in a thousand million. On the other hand, if we use Galileo's brilliant but approximate theory of the motion of a projectile to calculate the path of an actual cannonball, ignoring the highly significant factor of air resistance, then our calculated result and the experimental result could be widely different.

Indeed, by modern standards, it would be fair to say that Galileo's account of projectiles is so crude that we should count it as only a rough model.

Similarly, even Newton's theory of universal gravitation is now known to be no more than a brilliant approximation. Lagrange, writing in the days when scientists still believed that they were turning one by one the pages of the Book of Nature, wrote that 'Newton was the greatest genius that ever existed, and the most fortunate, for we cannot find more than once a system of the world to establish.' Not so, as Einstein demonstrated. Yet, Einstein's theories are also not the last word: quantum theory and relativity are inconsistent, and Einstein himself, proclaiming that 'God does not play dice!', rejected the basic reliance of quantum theory on chance events, and looked forward to a theory which would be deterministic. Recent experiments suggest that this view of Einstein's conflicts with his other deeply held beliefs about the nature of the physical universe. Certain it is that somewhere, beyond physicists' current horizons, are even more powerful theories of how the world is.

There is no sharp dividing line between scientific theories and models, and mathematics is used similarly in both. The important thing is to possess a delicate judgement of the accuracy of your model or theory. An apparently crude model can often be surprisingly effective, in which case its plain dress should not mislead. In contrast, some apparently very good models can be hiding dangerous weaknesses.

A topical example of the latter occurred a few years ago, when chemists who were building models of the earth's atmosphere failed to predict the hole in the ozone layer. Even when the computer models produced readings which – in retrospect – shouted out loud that something strange was happening, the scientists ignored them, preferring to believe that their models were already adequate to show any large scale phenomena in the atmosphere, and consequently that the 'ozone hole' data were freak readings, chance concurrences which meant nothing.

Eugene Wigner found the success of mathematics in explaining and predicting experience baffling. The ozone-layer example shows the mathematical modellers being baffled by the complexity and unpredictability of the world. In between these two extremes, mathematics continues to provide a wonderfully effective tool for making sense of the world around us.

Problem 10G *A falling raindrop*

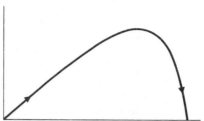

Fig. 10.9

A ball fired from a cannon is a very dense solid object travelling through a rather light medium (air). Nevertheless, the resistance of the air is considerable because, as experiment shows, it is roughly proportional to the square of the speed, and cannonballs travel fast. Consequently, the path of the cannonball will be foreshortened, as it were, as in Fig. 10.9, which is by no means a symmetrical parabola. Similar paths can be seen on television when a golf ball is driven from the tee.

In contrast, a raindrop falling through the air is a very light, non-rigid object travelling through a medium which is, relative to the raindrop, significantly dense. What factors would have to be taken into account in a sophisticated model of the fall of a raindrop?

Solutions to problems, Chapter 10

10A: *Olbers' paradox*

The perceived total brightness of all the stars in a spherical shell should be, according to this argument, *constant*, because the greater number of stars in more distant shells will exactly balance the loss of perceived brightness due to the greater distance. It follows that if there really are an infinite number of stars, and if they are uniformly distributed through space, then the entire sky should be uniformly bright. Indeed, it should be dazzling. Since this is not the case, it is plausible to conclude that the number of stars is finite, or that they are irregularly distributed, or that some other factor is at work.

10B: *The flying fish*

The fish leaves the water at an angle of 20°. Suppose that its velocity in that direction is v, then its horizontal velocity will be $v \cos 20°$ and its vertical velocity will be $v \sin 20°$. If the fish is flying through the air simply as a projectile, then its range will be

$$(2 \sin 20° \cos 20°)v^2/g,$$

and its maximum height will be

$$(\sin^2 20°)v^2/2g.$$

The ratio range/height will be $(4 \cos 20°)/\sin 20° = 11$ (approximately). But according to the figures of the problem, the ratio of range to height is $7/\frac{1}{2} = 14$. Therefore the fish is travelling more than 25% further, for the height reached, than it 'ought to do'. We conclude that it is benefiting from a gliding effect, presumably using its pectoral fins.

10C: *A dice simulation*

If each generation produces one son, the family will continue for ever, just. In this problem you have three chances of doing worse than that and only one chance of doing better, at each generation, which suggests that the family will die out. In fact, the Watson equation is:

$$x = \tfrac{1}{2} + \tfrac{1}{3}x + \tfrac{1}{6}x^2$$

Observing that $\tfrac{1}{2} + \tfrac{1}{3} + \tfrac{1}{6} = 1$, we see that this has the root $x = 1$, and this is the only root that fits the problem. (The other root is $x = 3$.) So it is certain that the family will eventually become extinct, as expected.

10D: *How many sources?*

In order to have exactly three soures, it is necessary and sufficient that the following independent conditions are all satisfied:

(1) The original river bifurcates: probability $\tfrac{1}{2}$.

(2) One tributary branches but the other does not: probability $\tfrac{1}{4} + \tfrac{1}{4} = \tfrac{1}{2}$.

(3) Both branches of the tributary that does branch do not branch further: probability $\tfrac{1}{4}$.

Since these events are independent, the probability that they all occur is the product of their probabilities:

$$\tfrac{1}{2} \times \tfrac{1}{2} \times \tfrac{1}{4} = \tfrac{1}{16}.$$

10E: *Where is the limit point?*

The interpretation of this solution is that the system is not stable, and instead of settling down to a steady state, or oscillating as in the previous example, will veer off to an extreme. The solution in this case is $(x, y) = (a/b, 0)$. The interpretation is that the predator population dies out through lack of prey.

10F: *Stabilizing pigs*

The simplest actions, which avoid the possible difficulty and embarrassment of actually telling farmers what they can and cannot do, are to:

- buy pigs when there is a glut on the market, which will help to raise prices and reduce the chance that farmers who have overproduced will be unable to sell their pigs and will consequently go bankrupt
- sell pork when there is a shortage, so that prices are brought down. The knowledge that prices will be limited by intervention will also discourage farmers from running to produce more pigs themselves in order to take advantage of high prices which will not last.

The result of these actions will be to counter the natural instability of the system. Note however that these moves could create their own difficulties. For example, if the prices offered to take pigs out of the market are too high, farmers may produce pigs simply to sell directly into store. As economists, and politicians, know very well, it is not easy to correct the natural instabilities in economic systems. (The Pig Marketing Board was, in fact, very successful right up to the start of the Second World War, when lack of feedstuffs in wartime conditions led to large reductions in the pig population.)

10G: *A falling raindrop*

The smaller a drop of water, the greater the ratio between the surface tension and the volume of the drop. Surface tension will keep a very small drop roughly spherical, but the size of raindrops that fall in a shower are large enough to vary considerably from a spherical shape as they fall: nor will they adopt one other fixed shape, but rather their shape will tend to oscillate. The force of gravity will pull them downwards, but electrostatic forces in the atmosphere must also be considered, and they can vary in direction. The smaller the raindrop, the greater the relative effect of electrostatic forces. The raindrop may acquire extra mass, or lose it, depending on the state of the atmosphere and the temperature. The resistance of the air will depend on the speed of the drop and the density of the air: there are no simple formulae for modelling the effects of air resistance, which will be relatively larger the smaller the drop.

Chapter 11
The enjoyment of mathematics

> There is no imaginable mental felicity more serenely pure than
> suspended happy absorption in a mathematical problem.

So wrote Christopher Morley in his novel *Plum Pudding*. He was the
son of a famous mathematician (Frank Morley), so he could have
been slightly biased. G. H. Hardy expressed a more popular view and
offered an explanation, when he referred in *A Mathematician's Apo-
logy* to:

> ... the puzzle columns of the popular newspapers. Nearly all their immense
> popularity is a tribute to the drawing power of rudimentary mathematics ...
> They know their business: what the public wants is a little intellectual 'kick',
> and nothing else has quite the kick of mathematics. The fact is that there are
> few more 'popular' subjects than mathematics. Most people have some
> appreciation of mathematics, just as most people can enjoy a pleasant tune.

In what does this enjoyment exist? It could start with an enjoyable
feeling of *wonder*: How extraordinary! Who would have thought
that! Or it might start with a closely related emotion of *surprise*. Is
that really so? You don't say! Who would have thought it!

Paradox also gives us a thrill. Come off it – that's not possible!
Don't be absurd! *Mystery* is found as much in mathematics as in
detective stories. Indeed, the mathematician could well be described
as a detective, brilliantly exploiting a few initial clues to solve the
problem and reveal its innermost secrets. An especially mathematical
mystery is that you can often search for some mathematical object,
and actually know a lot about it, *if it exists*, only to discover that in
fact it does not exist at all – you knew a lot about something which
cannot be.

What happens when the surprise is explained? When the paradox is
resolved? When the mystery is revealed? Mathematicians get a different kind of pleasure from the *illumination* of solving a problem, when
what was once mysterious and obscure is made plain. Revealing the
hidden connections in a situation is delightful – like reaching the top
of a mountain after a hard climb, and seeing the landscape spread out
before you. All of a sudden, everything is clear! If the result is

extremely simple, so much the better. To start with confusing complexity and transform it into revealing *simplicity* is a marvellous reward for hard work. It really does give the mathematician a 'kick'!

Puzzle composers deliberately exploit the delight in mystery and surprise by the manner in which they design and present their posers, as Henry van Etten admitted in his *Mathematical Recreations* (1633). Explaining that he had given answers to most of the problems, without 'speculative demonstration' which he left to the mathemat-

---

Problem 11A *Langley's adventitious angles*

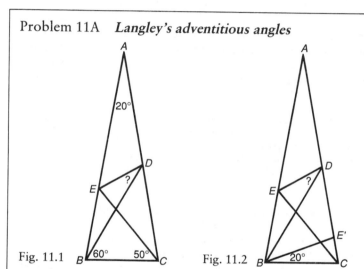

Fig. 11.1   Fig. 11.2

This problem, named after its first proposer, E. M. Langley, is famous for being harder than it looks. The isosceles triangle *ABC* has a vertex angle of 20°, so that each of its base angles is 80° (see Fig. 11.1). The two lines *EC* and *DB* have been drawn to make the marked angles with the base of the triangle and the problem is to find the size of angle $\widehat{BDE}$. In elementary geometry textbooks such problems are usually solved, rather easily, by 'angle chasing': calculating as many angles as possible in sequence, until the size of the required angle is found. In the present case this does not work, as you can verify for yourself. However, angle chasing does work *after* we have added a single construction line, as in Fig. 11.2. It is almost paradoxical that (as so often in geometry), by making the figure slightly more complicated, the solution of the problem becomes much simpler.

What is the size of angle $\widehat{BDE}$, by angle chasing?

---

icians among his readers, he wrote: '[these things] ought to be concealed as much as they may ... for that which doth ravish the spirit is an admirable effect whose cause is unknown, which if it were discovered, halfe the pleasure is lost.' Van Etten's conclusion that 'halfe the pleasure' is lost shows a nice judgement. A certain pleasure is lost when a mystery is revealed, and yet most mathematical mysteries naturally reveal so many new and elegant ideas, ingenious explanations, not forgetting new mysteries and further questions that the mathematician is usually more than happy to discover the answer to his or her initial questions.

Puzzle composers share another feature with mathematicians. They know that, generally speaking, the simpler a puzzle is to express, the more attractive it is likely to be found: similarly, simplicity is for both a desirable feature of the solution. Especially satisfying solutions are often described as 'elegant', a word that – no surprise here – is also used by scientists, engineers and designers, indeed by anyone with a problem to solve. However, simplicity is by no means the only reward of success. Far from it! Mathematicians (and scientists and others) can reasonably expect two further returns: they are (in no particular order) firstly the power to do things, and secondly the perception of connections which were never before suspected, leading in turn to the insight and illumination that mathematicans expect from their best arguments.

The power that successful solutions confer on the solver is experienced by pure and applied mathematicians alike. We so take for granted the power of science and technology to send men to the moon, or splice genes, or split atoms, or speculate about the first fractions of a second in the existence of the universe, that we can forget that these achievements are dependent on mathematical discoveries. Jacques Ozanam, writing three centuries ago in his *Cursus Mathematicus*, was less blasé about the achievements of the mathematics of his day. This is his encomium on the humble trigonometry that we all started to tackle at school:

'Tis by Trigonometry only, that the Courses of the Phaenomenas and Changes which happen in the Universe, can with any certainty, be discover'd; ... nor can any one arrive at the knowledge of the Motions of the Coelestial Bodies, but by that of the most simple Figures, which are Triangles, whose Mensuration is taught by Trigonometry. By whose assistance Engineers measure accessible and inaccessible Distances ... Astronomers the Distances of Stars ... the Art of navigation depends entirely on Trigonometry ... The Usefulness of Trigonometry is so great, that it is in a manner impossible to live without it ...

As indeed, it still is, except that we take it for granted. Spotting connections, observing similarity in apparent difference, realizing unity in diversity, seeing the harmonious manner in which the various parts of mathematics fit together, is a different kind of delight, which depends, perhaps, on our deepest feelings about the world, which so often seems to be chaotic, without purpose and meaningless, while also displaying pattern and order.

Problem 11B *Prince Rupert's cube, and the puzzle of the divided square*

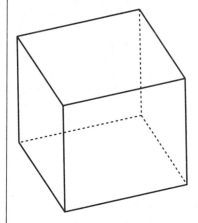 

Fig. 11.3a   Fig. 11.3b

Prince Rupert was a nephew of King Charles I who fought in the Civil War and was also an early member of the Royal Society. He asked this extremely simple question: 'What is the largest cube that can be passed through a square hole cut in a given cube?' (See Fig 11.3a.)

Figure 11.3b illustrates a very popular Victorian puzzle, which has the same virtue of simplicity although, being a composed puzzle presented in an artificial setting, it is not as simple as the pure mathematician might make it: 'A charitable individual built a house in one corner of a square plot of ground, and let it to four persons. In the ground were four cherry trees, and it was necessary so to divide it, that each person might have a tree and an equal portion of garden ground. How is to be divided?'

These are the motivations which get the mathematician going, and the rewards which success procures. What happens in between? The mathematician gets both benefits, at the same time! The problem starts with questions. You try to make sense of them, you spot a connection, a pattern, it works, no, it doesn't work, but it suggests something else, the solution you are looking for has this property, so it ought to . . . ah! now you're getting somewhere . . . And so the process continues, surprises, mysteries, connections, flashes of simplicity, all mixed up . . . , until the solution is found, the clouds disperse, and at last you can see clearly. As so often, the journey can be just as enjoyable as the moment of arrival.

These delightful features relate to all three of the aspects of mathematics that we have discussed in this book. Mathematics-as-science naturally starts with mysterious phenomena to be explained, and leads (if you are successful) to powerful and harmonious patterns. Mathematics-as-a-game not only starts with simple objects and rules, but involves all the attractions of games like chess: neat tactics, deep strategy, beautiful combinations, elegant and surprising ideas. Mathematics-as-perception displays the beauty and mystery of art in parallel with the delight of illumination, and the satisfaction of feeling that *now* you understand.

To achieve such satisfaction, however, may take a great deal of effort, as Alison Hird, a sixth form pupil, explained in a tape-recorded discussion:

Looking at a painting is very superficial, it is a subjective rather than an objective [view], and if you get into maths it is far more objective . . . the further you get into it, the more fascinating it can become, but a painting can be beautiful without really understanding the artist's meanings and emotions . . . whereas, with a formula you have to go deep into it . . . and that provides a far clearer view of beauty, if you like, than just looking at a painting and saying 'That's lovely!'.

The pleasure of surprise

Topology is sometimes called 'rubber-sheet geometry' because it is concerned only with properties that are unchanged when the figure is stretched like a sheet of rubber (but without being torn). Lengths and angles are of no consequence in topology, which suggests that in some sense it is very simple. Indeed it is, in one sense, and yet our very familiarity with angles and lines, and our reliance on them, ensures that many simple topological questions are difficult to answer, because our intuition is not very helpful. For example, if you take a

---

Problem 11C   *Julius Caesar's breath*

Sir James Jeans, the astronomer, asked this question: 'What is the chance that with your next breath you will breathe in at least one molecule which was once breathed out by Julius Caesar?'

---

left-hand glove and turn it inside out, do you get another left-hand glove, or a right-hand glove?

Figure 11.4 depicts a very flexible but solid shape made, let's imagine, of a soft kind of plasticine. This plasticine can be stretched and squashed as much as we like, but we are not allowed to cut it or tear it. The puzzle is to separate the two circular loops, and it is quite obviously impossible. Except that it isn't! By following the sequence of moves shown in Fig. 11.5, the loops do separate in a continuous transformation. Such tricks are amazing because, like the best magical tricks, they confound our confident expectations.

Fig. 11.4

Fig. 11.5

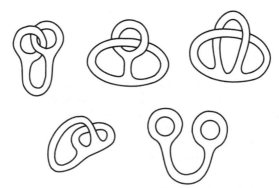

Problem 11D *The handcuff*

Fig. 11.6

The loop is clearly handcuffed by the shape on the left. How can it be continuously transformed, without breaking or tearing, so that *one* of the loops of the handcuff comes off?

Problem 11E *How many darts?*

Fig. 11.7

Figure 11.7 shows a standard dartboard. You throw six darts at the board, completely at random, except that you are accurate enough to get them all actually on the board. You do not score any bullseyes, and doubles and trebles do not count (so that '7', 'double 7' and 'treble 7' all count the same). What is the chance that two of them will stick in the same segment?

304 • YOU ARE A MATHEMATICIAN

Problem 11F   *Smith and Watson's scores*

Suggest some plausible batting scores for Smith and Watson such that the situation described in the text could arise.

Paradoxical pleasures

If topology takes the prize for upsetting mathematicians' intuitions, probability and statistics are not far behind. Suppose that you are at a party. There are twenty-three people in the room. Someone starts to tell astrological fortunes so you are soon revealing your dates of birth. What is the chance that two people in the room have the same birthday?

There are 365 possible birth dates in the year, not counting leap years, and there are only 23 people present, so the probability might seem to be quite small, but this first impression is false. Calculation shows that it is greater than fifty–fifty!

The batting paradox

Smith and Watson are keen rivals for the club batting trophy. In the first half of the season, Smith's average is comfortably ahead of Watson's, so Smith starts the second half feeling confident of winning the cup. Sure enough, his average during the second half of the season is also better than Watson's, but, to his chagrin, when the averages for the whole season are calculated – Watson wins! (See Problem 11F.)

As it happens, this is more than just a puzzle, because the same result can follow in much more serious situations. Suppose that you are a scientist comparing the number of descendants produced in a litter by different breeds of rabbits. The two breeds you are comparing are the Softies and the Streakies, and your conjecture is that the Softies produce more progeny.

In your first experiment, your conjecture is confirmed, because the average number of progeny, in a birth, of the Softies, is higher than that of the Streakies. The second experiment confirms the hypothesis. You are of course, delighted, but as you prepare to publish your results you realize to your distress that the overall figures *contradict* your hypothesis.

How is it possible for two experiments, each of which *genuinely* supports a scientific hypothesis, to combine to destroy it?

Problem 11G *Two paradoxes of more than and less than*

The idea of infinity has often confused mathematicians. Galileo noticed this paradoxical situation. Write the integers in sequence, and underneath write the sequence of their squares.

1	2	3	4	5	6	7	8	9	10	11	12	13	⋯
1	4	9	16	25	36	49	64	81	100	121	144	169	⋯

Galileo reasoned that, on the one hand, there are clearly exactly as many squares as there are integers, because there is just one square for each integer and one integer for each square. On the other hand, it is just as clear that there are *more* integers than squares, because most integers are *not* squares. Paradox! What is the resolution of this paradox?

The very idea of negative numbers also confused many seventeenth-century mathematicians. Arnauld, a friend of the philosopher and mathematician Pascal, argued that if the number 'minus one' existed, then

$$-1/1 = 1/-1$$

But, he argued, this is absurd: -1 is less than 1, so here you would have a lesser number divided by a larger number, equal to the same larger number divided by the lesser number: as if 4/13 could possibly equal 13/4. Therefore, Arnauld concluded, negative numbers don't exist. Where is the flaw in Arnauld's argument?

The mystery of the odd perfect numbers

Arnauld thought (see Problem 11G) that he had disproved the existence of negative numbers by arguing that, *if* they existed, *then* they would have a certain (absurd) property. Ironically, mathematicians often infer all sorts of properties about objects which they only suspect, or hope, actually exist. If their suspicions turn out to be unfounded, then they seem to end up by knowing rather a lot about something which does not exist and which might seem, therefore, not to have any properties at all.

This curious state of affairs is illustrated by the problem of *odd perfect numbers*. A perfect number is equal to the sum of all its divisors, including 1 but excluding, of course, the number itself. The

smallest is 6, because $6 = 1 + 2 + 3$. The next is $28 = 1 + 2 + 4 +$
$7 + 14$, and the next two are 496 and 8128. It is no coincidence that

$$6 = 2(2^2 - 1), \qquad 28 = 2^2(2^3 - 1),$$

$$496 = 2^4(2^5 - 1), \qquad 8128 = 2^6(2^7 - 1).$$

The general pattern for all *even* perfect numbers, which Euclid
proved in his *Elements*, is that they are a prime number in the form
$2^n - 1$ multiplied by 2^{n-1}. Odd perfect numbers, however, do not fit
Euclid's formula – if they exist at all. It is known that, if they exist,
then they are very large, greater than 10^{200}. An odd perfect number
must have at least eight distinct prime factors (eleven if it is not
divisible by 3) and must be divisible by a power of a prime greater
than 10^{18}. The greatest prime factor must be greater than 300,000 and
the second must be greater than 1000. Any odd perfect number less
than 10^{9118} must be divisible by the 6th power of a prime.

No doubt there are many other properties of odd perfect numbers
waiting to be discovered, and maybe one day an actual odd perfect
number will be found – with all these properties. But perhaps it
won't!

Illumination: revealing hidden connections

One of the greatest delights of mathematics is finding unity among
apparent diversity; realizing that situations (or objects) that seemed to
be quite different are actually basically the same. Spotting unexpected
connections is not only a source of pleasure, it is an essential step in
the development of mathematics. Without making connections, mathematics would quickly turn into a collection of separate topics,
studied by mathematicians who had nothing to say to colleagues
outside their own speciality. Some mathematicians do bemoan the
fact, as they see it, that mathematics is heading in that direction. Yet
all the time, links are being made.

Fermat and Descartes made one of the grandest connections in the
history of mathematics, by linking elementary algebra and geometry.
It turned out that statements in algebra could be represented as
statements about geometrical figures, and that conversely geometrical
figures, including the many named curves which had been studied
from the Greeks onwards, could be interpreted as algebraic equations.
The result of this linkage was that geometrical insight and visualiza-
tion was brought to the study of algebraic problems, while – conversely – algebraic ideas illuminated geometrical situations. We have

seen examples of such insights in the way that the properties of the ellipse and hyperbola match the difference (which is very small – it's only a change of sign!) in their algebraic equations.

In Chapter 3 we saw that there is a surprising connection between the fact that $n^2 + n + 41$ is prime for all values of n from zero to 40, and the fact that $e^{\pi/163}$ is almost exactly an integer. This is not a connection between two great branches of mathematics, but it is nevertheless very subtle and cunning. To make it, one has to delve very deeply into modern number theory, like a geologist who observes that two rock formations many miles apart, and apparently unrelated, are in fact connected by the hidden pattern of strata in between.

It is no surprise that these investigations, into what is called 'class number theory', have led to many other powerful results. To continue with our geological analogy, we would not be surprised if the geologist, in making such a distant connection by a deep study of the intervening strata, should discover valuable deposits, or an oil reservoir or two, in between.

The honour of making such profound and wide-ranging connections is given only to a fortunate few mathematicians, but we can all enjoy getting a kick out of making lesser connections. Here is an example. Figure 11.8 shows an equilateral triangle, and a point inside it. The problem is simple to state: if we know the distance of the point from the three vertices, what is the size of the triangle edge? The motivation for this little problem is curiosity, plus the knowledge that solving such problems often turns up something interesting, plus, perhaps, the challenge of finding the connection in the form of algebraic equation as quickly and simply as possible.

Supposing that we have solved that challenge, we eventually get to this equation:

$$3(a^4 + b^4 + c^4 + d^4) = (a^2 + b^2 + c^2 + d^2)^2.$$

Fig. 11.8

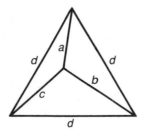

Was the effort worth it? Yes, it seems it was, because the final equation is not only symmetrical in *a*, *b* and *c*, which was expected, but it is symmetrical in *d* as well! Apparently, we could have started just as well with a point distance *a*, *c* and *d* from the vertices of an equilateral triangle whose edge is *b*. In Fig. 11.9, the individual figures for the other three cases are given. They invite the question, 'Where does this symmetry come from?' Noticing that each figure is composed of two triangles from the original figure, joined edge to edge, suggests that these figures do in fact fit together, to make a larger figure that has many symmetries (Fig. 11.10). With a bit of trial

Fig. 11.9

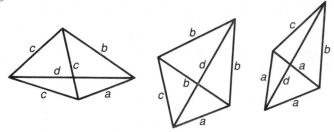

Fig. 11.10

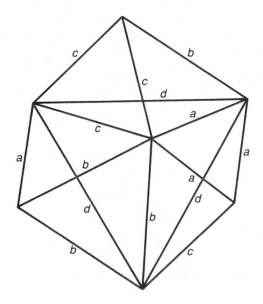

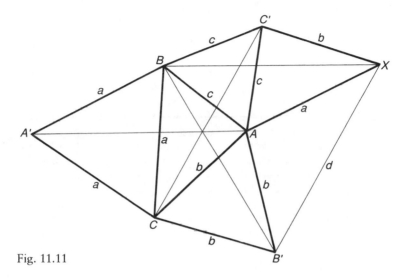

Fig. 11.11

and error, this is it. It is a part of an infinite tessellation in which the edges of the four different lengths appear symmetrically.

At this point, our small adventure might come to an end. It doesn't, because we notice something that depends entirely on our own past experience. (How typical of mathematics!)

The figure we have arrived at (Fig. 11.11) is remarkably similar to the diagram which shows how the Fermat point of a triangle, such as *ABC*, can be constructed, by drawing equilateral triangles on each edge, and then joining *AA'*, *BB'* and *CC'*. These lines, which meet at angles of 60° and are also all of the same length, concur in the Fermat point.

We see this Fermat figure on the left side of our diagram. On the right we have added the point *X*, joined as shown to *C'*, *A*, *B'* and *B*. The line *BX* is parallel to *A'A* and *B'X* is parallel to *CC'*, and *XBB'* is equilateral because of the properties already mentioned. Its side is *d*, the remaining length in the equation, and the point *A* inside *XBB'* is at distances of *a*, *b* and *c* from the vertices. Once again, we see that problems and situations which may appear at first sight to be quite distinct are actually close relatives.

Problem 11H *Simplicity in calculation*

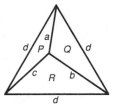

Fig. 11.12

How should we go about calculating the algebraic relationship between a, b, c and d, as simply as possible? We could, for example, use Heron's formula for the area of triangle, given its sides, to write down the areas of the three triangles P, Q and R, and then equate their sum to the area of the original equilateral triangle. According to Heron's formula, the area of a triangle with sides x, y and z is $\sqrt{[(s(s-x)(s-y)(s-z)]}$ where $s = \frac{1}{2}(x + y + z)$. It can also be written as

$$\tfrac{1}{4}\sqrt{[(x + y + z)(-x + y + z)(x - y + z)(x + y - z)]}.$$

Alternatively, we could add a construction line, as in the left-hand triangle in Fig. 11.13, where h is the length on an altitude. We then find equations connecting a, b, c, d and h, and eliminate h to find the relation we want. A third possibility (and of course there are yet others) is to label the angles at the centre, say, a, β and γ, and use the cosine formula which says, for example, that

$$d^2 = b^2 + c^2 - 2bc \cos a.$$

If we can find a relationship between the cosines of a, β and γ (on the grounds that $a + \beta + \gamma = 360°$), then once again we could be on our way.

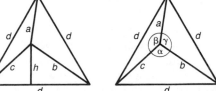

Fig. 11.13

In theory, any of these approaches will work. They might work well in practice also if we had the use of a powerful computer-algebra program to do all the heavy manipulative work for us. By hand, however, they are all difficult, but not equally so.

Which is the simplest?

Illumination, certainty and proofs

Mathematicians find immense satisfaction in proofs: but what kind of proofs? Which offers most delight, a proof that offers certainty, or a proof that gives illumination? Let's look at a simple example.

Consider the fact that $\sqrt{2}$ is not a rational number. This can never be proved scientifically because, however many fractions you test, searching for one whose square is 2 (and always failing), you will never be certain that there is not another fraction, which you have not tried, that will work.

Using the game of algebra is much more promising; in fact, this is just what traditional proofs of this theorem do. Suppose that $\sqrt{2} = p/q$, where p and q are integers and p/q has been 'reduced to its lowest terms', as the schoolbooks say, so that p and q have no common factor. We have

$$2 = p^2/q^2 \quad \text{and so} \quad p^2 = 2q^2.$$

Therefore p^2 is even, and so p is even also. Write $p = 2r$. Then

$$(2r)^2 = 2q^2 \quad \text{and so} \quad 2r^2 = q^2.$$

By the same argument as before, q must be even also. But in that case p and q *do have* a common factor, 2, which contradicts our original assumption. This contradiction proves that $\sqrt{2}$ is not rational.

That proof is perfectly sound, but it is not very illuminating. The contradiction appears, but almost by magic: why was it there in the first place? We do not know.

Problem 11I   *Roughly root 2*

Although $\sqrt{2}$ is not a fraction, you could discover a lot of interesting facts during an unsuccessful search for a fraction that equals $\sqrt{2}$, for example, the fact that the fractions in this sequence approach $\sqrt{2}$ very closely:

$$\left(\tfrac{3}{2}\right)^2 = 2\tfrac{1}{4} \quad \text{and} \quad 3^2 = 2 \times 2^2 + 1,$$

$$\left(\tfrac{7}{5}\right)^2 = 1\tfrac{96}{100} \quad \text{and} \quad 7^2 = 2 \times 5^2 - 1,$$

$$\left(\tfrac{17}{12}\right)^2 = 2\tfrac{1}{144} \quad \text{and} \quad 17^2 = 2 \times 12^2 + 1,$$

$$\vdots$$

How does this sequence continue?

The next proof is by looking at what is happening in a special way. This proof is *illuminating*. Suppose that p and q are two integers with no common factor. We now observe that when they are squared, they will still have no common factor. Therefore no fraction p^2/q^2 can cancel down. In particular, p^2/q^2 certainly cannot cancel down to equal 2. Therefore, $\sqrt{2}$ cannot be a fraction.

Now we can see clearly the principle behind the fact that $\sqrt{2}$ is irrational. It all boils down to the fact that a fraction that does not cancel will still not cancel when it is squared. (Actually, there is a gap in our argument at this point, which we will gloss over here.) Put like that it is not only completely obvious that $\sqrt{2}$ cannot be a fraction: it is just as obvious that $\sqrt{3}$ isn't, $\sqrt{5}$ isn't, in fact *no* square root of an integer is a fraction, unless the integer is already the square of an integer, in which case its square root is also an integer.

We can even claim another bonus: this argument is not limited to square roots. If you cube a fraction which doesn't cancel, it still won't cancel. So the cube root of 2 is irrational, and so are the cube roots of all numbers which are not perfect cubes, and the same for fourth roots and fifth roots, and so on. Amazing! By one illuminating argument we have solved our original problem and a host of other related problems.

Individual differences and changes in taste

Which examples of mathematics have you enjoyed most? The chances are that, whatever your preferences, many other readers will beg to differ. Mathematicians are by no means agreed on what mathematics is attractive, or worthwhile. On 19 November 1989, at the University of Tokyo, Yasumasa Kanada and Yoshiaki Tamura used a Hitac S-820 80E computer to calculate π to 1,073,740,000 decimal places. They used some advanced mathematics to write the computer program, and they were in a long line of mathematicians who have researched more and more powerful methods of calculating π accurately. Yet many mathematicians would say that this is 'not mathematics', but a not-very-interesting exercise in using a supercomputer. What do you think?

While the Indian prodigy Ramanujan was still at high school, he discovered that

$$e^{i\pi} = -1$$

and was later disappointed to learn that Euler had found the same result in 1743. At that time, this formula was considered an extraordinary and beautiful connection between π, the ratio of the circum-

Problem 11J   *Prime factorizations*

Consider numbers of the form $10^n - 1$. Many of them have been factorized because their special form makes factorization *relatively* easy. Many of their factors also display marked patterns, such as

$$M_{38} = 10^{38} - 1 = 3 \times 3 \times 11 \times 909090909090909091$$
$$\times 11111111111111111111,$$
$$M_{39} = 10^{39} - 1 = 3 \times 3 \times 3 \times 37 \times 53 \times 79$$
$$\times 265371653 \times 900900900900990990990991.$$

Numbers of the form $10^n + 1$ have also been extensively studied. This is one factorization:

$$10^{60} + 1 = 73 \times 137 \times 1676321 \times 99990001$$
$$\times 5964848081 \times 100009999999899989999000000010001.$$

If these patterns intrigue you, you may be attracted to the problem of factorizing $M_6 = 10^6 - 1 = 999999$,   and $M_{12} = 10^{12} - 1 = 999999999999$.

ference of a circle to its diameter, i, the square root of -1, and e, which is the base of natural logarithms. The equation was short, simple, compact, very surprising, and mysterious, at a time when mathematicians did not clearly understand the nature of complex numbers.

Two centuries later, most professional mathematicians still find this formula beautiful, but there is some dissent. The reasons for the formula are now well understood, even by undergraduates, so the mystery and surprise have largely vanished, leaving only the simplicity and elegance behind. Familiarity has bred, not contempt, but calm acceptance.

Stirling's formula, named after James Stirling (1692–1770), is an approximate formula for calculating $n!$ (pronounced 'n factorial', or 'n bang'). The integer $n!$ is the product of the first n integers, so $n!$ increases in size very quickly indeed, as shown in Table 11.1 for $n = 1$ to 12.

Table 11.1

n	$n!$	n	$n!$	n	$n!$	n	$n!$
1	1	4	24	7	5,040	10	3,628,800
2	2	5	120	8	40,320	11	39,916,800
3	3	6	720	9	362,880	12	479,001,600

Stirling's formula states that

$$n! \approx n^{n+1/2}e^{-n}\sqrt{(2\pi)}$$

(where the sign $\approx$ means approximate equality), so it involves π and e, but not i. Speaking (of course) personally, I find Stirling's formula more attractive than Euler's, perhaps beause I do not understand it so well, and therefore find it more mysterious.

Because I like Stirling's formula, I am inclined to think that others will like it also, but that assumption is, of course, naive. Here is what a great modern mathematician, the Russian V. I. Arnol'd, said in an interview in 1987: 'By the way, the 200 year interval from Huygens and Newton to Riemann and Poincaré seems to me to be a mathematical desert filled only with calculation.'

Well, that would seem to dispose of Stirling and a great many other brilliant mathematicians, and all their achievements! Talking of calculation – but we mean the kind of calculation that depends on deep understanding, imagination, and insight, and involves brilliant mathematical combinations – here are some equations which involve the number π, in very different ways, to add to the series on pages 88 and 92. How would you rate them for elegance and beauty, for surprise and mystery? (Euler used the last formula to calculate the value of π to 20 places in one hour.)

Vieta (1540–1603)

$$\frac{2}{\pi} = \sqrt{\frac{1}{2}}\sqrt{\frac{1}{2} + \frac{1}{2}\sqrt{\frac{1}{2}}}\sqrt{\frac{1}{2} + \frac{1}{2}\sqrt{\frac{1}{2} + \frac{1}{2}\sqrt{\frac{1}{2}}}}\cdots$$

Wallis (1616–1703)

$$\frac{2 \times 2 \times 4 \times 4 \times 6 \times 6 \times 8 \times 8 \times 10 \times 10 \times \cdots}{1 \times 3 \times 3 \times 5 \times 5 \times 7 \times 7 \times 9 \times \;\; 9 \times 11 \times \;\cdots}$$

Leibniz (1646–1716)

$$\frac{\pi}{4} = 1 - \frac{1}{3} + \frac{1}{5} - \frac{1}{7} + \frac{1}{9} - \frac{1}{11} + \frac{1}{13} - \frac{1}{15} + \frac{1}{17} - \cdots$$

Euler

$$\frac{\pi}{4} = \frac{7}{10}\left[1 + \frac{2}{3}\left(\frac{2}{100}\right) + \frac{2 \times 4}{3 \times 5}\left(\frac{2}{100}\right)^2 + \cdots\right]$$

$$+ \frac{7584}{10^5}\left[1 + \frac{2}{3}\left(\frac{144}{10^5}\right) + \frac{2 \times 4}{3 \times 5}\left(\frac{144}{10^5}\right)^2 + \cdots\right].$$

Problem 11K   *How do they rate?*

How do you rate these ten mathematical theorems for their attractiveness? The ratings given by seven mathematics teachers are given in the Solutions, together with the average ratings, for the last seven theorems only, of the seventy or so professional mathematicians who responded to a quiz which included these theorems among others, in the journal *Mathematical Intelligencer* ('Which is the most beautiful?', vol. 10(4), 1988, and 'Are these the most beautiful?', vol. 12(3), 1990. The exact wording in the journal quiz was sometimes very slightly different.)

A. $142857 \times 3 = 428571$.

B. If you join each vertex of a triangle to the middle of the opposite side with a straight line, all the three lines will go through the same point.

C. $1 + 3 + 5 + 7 + 9 + 11 = 6^2$.

D. There are an infinite number of prime numbers.

E. There are five regular polyhedra.

F. $1 + 1/2^2 + 1/3^2 + 1/4^2 + 1/5^2 + 1/6^2 + \cdots = \pi^2/6$.

G. There is no fraction whose square is 2.

H. (The four-colour theorem): every plane map can be coloured with four colours.

I. It is impossible to find three crossing-points on a square grid which are exactly the vertices of an equilateral triangle.

J. At any party, there is a pair of people who have the same number of friends present.

The view expressed by Arnol'd, an extreme view from an extremely brilliant mathematician, is an expression of his own personal style and preferences. Mathematicians vary in their attitude to calculation as they do in their attitudes to structure and pattern. The difference is very well put by the British chess grandmaster Jon Speelman discussing the style of Nigel Short:

> Short is essentially a positional player with a tremendous natural feel for structure and harmony that on occasion allows him to cut through to the very

heart of a complex position. Calculation comes less naturally to him, and although he is very good at it I believe that he finds it somewhat of a chore, and certainly doesn't enjoy calculation for its own sake as I do.

It is no surprise that there is a close analogy here between chess and mathematics, or that Short more resembles Arnol'd while Speelman resembles Euler or other great calculators.

Did you know that at any party there are bound to be at least two people who have the same number of friends present? Does this mathematical fact appeal to you? When more than seventy professional mathematicians scored this mathematical theorem for beauty, from 10 for really beautiful, to zero for not at all, the average score was just over $4\frac{1}{2}$, but individual scores varied from zero to 8.

When seven teachers in a conference audience were asked to rate the same fact in the same way, they gave the scores 2–7–8–8–8–7–10, and there were gasps in the hall as the highest and lowest scores were announced.

Clearly, as Reuben Hersh, co-author of *The Mathematical Experience,* observes, such judgements in mathematics vary widely:

> There is an amazingly high consensus in mathematics as to what is 'correct' or 'accepted'. But alongside this, and equally important, is the issue of what is 'interesting' or 'important' or 'deep' or 'elegant'. These aesthetic or artistic criteria vary widely, from person to person, speciality to speciality, decade to decade. They are perhaps no more objective than aesthetic judgements in art or music.

Changing historical taste

We have already met the *nine-point-circle* theorem: the midpoints of the sides, the feet of the altitudes, and the midpoints of the lines joining the vertices to the centre of gravity all lie on a circle. This was considered beautiful for many years. It was surprising, and elegant, and simple. As time passed, however, more properties of the circle were discovered. Feuerbach (1800–1834), after whom the circle is often named, proved that the incircle and the three excircles all touch the nine-point circle. Their points of contact made four more 'special points' on the circle. Other mathematicians added more and more points, until there were more than forty points on the circle . . . but by then most mathematicians had moved on, new geometries were holding their attention, and the nine-point theorem drifted into a backwater.

The story and changing fashions for mathematical facts and ideas, emphasizes the personal equation: what matters is what attracts *you*, what surprises *you*, what you find simple and elegant, what challenges you, and this will depend on your past experience, as well as on your feeling for the challenge of the new.

Romantic versus classical

Some individual differences in appreciation can be linked to differences in temperament. Some people naturally prefer the calm and ordered world of classical art to the emotion and disorder of romanticism. Francois Le Lionnais found the same contrast in mathematics: 'By contrast with classical mathematical beauty we are now going to examine another sort of beauty which can be described as romantic. Its underlying principle is the glorification of violent emotion, nonconformism and eccentricity.' Leaving aside the question of whether his characterization of 'romantic' is itself exaggerated and eccentric, we reproduce in Fig. 11.14 the kind of illustration he presented of romantic mathematics: it represents the solution to a differential equation.

Fig. 11.14

The classical spirit is well represented in mathematics by Bourbaki, a group of mathematicians, originally all French, who took their name from a nineteenth-century French general. Their monumental series of volumes aimed to present the whole of modern mathematics in an ordered and logical manner, starting always from first principles. The result was textbooks which are extremely abstract, and difficult to read. Their programme worked well enough as long as they were tackling those parts of mathematics which are well established – mainstream as it were – but worked less well when they got to the idiosyncratic and less-developed portions of mathematics.

Nevertheless, the Bourbaki approach became tremendously fashionable among young mathematicians – mathematics is subject to fashion like everything else! – and this prompted Freeman Dyson, who started life as a brilliant mathematician and became a world-famous physicist, to praise 'unfashionable mathematics' as follows:

Unfashionable mathematics is mainly concerned with things of accidental beauty, special functions, particular number fields, exceptional algebras, sporadic finite groups. It is among these unorganized and undisciplined parts of mathematics that I would advise you to look for the next revolution in physics. They have a quality of strangeness, of unexpectedness. They do not fit easily into the smooth logical structures of Bourbaki. Just for that reason we should cherish and cultivate them, remembering the words of Francis Bacon, 'There is no excellent beauty that hath not some strangeness in the proportion.'

Dyson continues, commenting on the 'sporadic finite groups': 'The only thing these various discoveries had in common was a concrete, empirical, experimental, accidental quality, directly antithetical to the spirit of Bourbaki.' The contrast could not be clearer. Where do you stand? Which kind of mathematics do you prefer?

On the walls of the physics department of Göttingen University, where many of the greatest mathematicians as well as physicists have worked, is the motto *Simplex Sigillum Veri*: 'Simplicity is the Sign of Truth'. Simplicity has always proved attractive – but how simple? Cannot a very complicated picture or situation also be simple, if you can see a strong pattern in it?

The mathematician and physicist Roger Penrose once asked which was more beautiful, a simple square grid, on the left, or the non-periodic pattern on the right of Fig. 11.15. He decided that he preferred the more complex pattern on the right: the square grid was just too simple. However, he may have been biased in his judgement, since it was Penrose who discovered the pentagonal tiling!

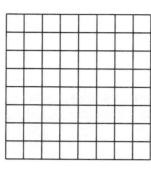

Fig. 11.15

Fig. 11.16

Figure 11.16 shows another contrasting pair: on the left, a portion of the Mandelbrot set of the function

$$f(z) = z^2 + c.$$

It swirls, it turns, it repeats itself, it is never still. On the right is the polyhedron formed by five regular tetrahedra inscribed in a regular dodecahedron. This image is static, simple, perhaps more classical?

Symmetry, and the beauty of asymmetry

Here is a scientist talking about symmetry: 'Until recently, traditional scientific thinking held that ultimately everything in the universe has a symmetrical equal and opposite counterpart, . . . but a beautiful theory has led to the finding that universal symmetry is sometimes slightly violated, and scientists believe they have come upon a truth even more beautiful than the simple perfection of symmetry.'

Mathematicians enjoy symmetry, but lack of symmetry has its appeal also. Here are two images: Fig. 11.17 shows a simple kind of cellular automaton investigated by Stanislaw Ulam, a mathematician

Fig. 11.17

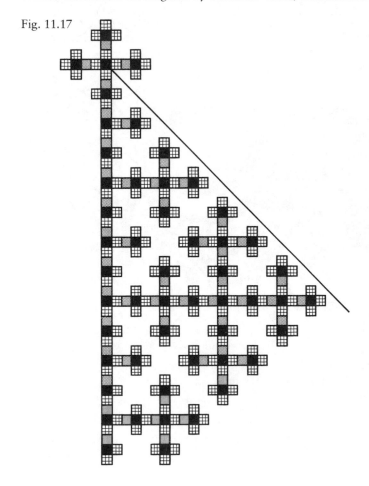

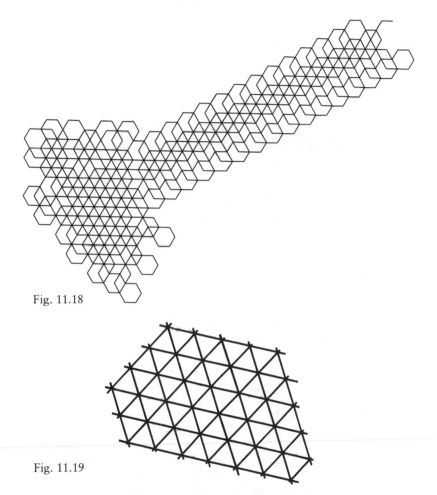

Fig. 11.18

Fig. 11.19

who worked on the construction of the first atomic bombs, and on early computers. Figure 11.18 shows a trace made by a programmed 'worm'. To create such patterns, imagine a worm which eats its way along the edges of an infinite tessellation of equilateral triangles, as in Fig. 11.19. There is a set of rules which tells the worm, every time it reaches an intersection, which way to turn, depending on the edges at that vertex, if any, which have already been eaten. Depending on which set of rules is chosen – and there are more than a thousand possible sets of rules – a different pattern will be created.

The first of the images above is symmetrical, and will always remain so. The second is far from being random, but has a weird kind of semi-symmetry. It also has the unusual feature of repeating to infinity in the northeasterly direction, after initially wandering around. Which is the most appealing to you, and why?

According to Le Lionnais (1901–84), 'After the paradoxes come the anomalies, the irregularities, indeed the monstrosities. They arouse some people's indignation and to others bring delight.' Indeed they do – Charles Hermite (1822–1901) wrote in a letter, 'I turn aside with a shudder of horror from this lamentable plague of functions with no derivatives.' (Functions with derivatives can be pictured as smooth curves. A function with no derivative is not smooth anywhere.)

We have already seen a geometrical example of the weird and monstrous curves to which Hermite was referring – the 'pathological' snowflake curve (page 73) which turns out to be not so pathological after all: self-similar fractal curves are now seen to be very common indeed, and very important. Here is an arithmetical, or algebraical, example of a 'pathological' series:

$$1 - 1 + 1 - 1 + 1 - 1 + 1 - 1 + 1 - 1 + 1 - \cdots .$$

If the terms are grouped like this:

$$(1 - 1) + (1 - 1) + (1 - 1) + (1 - 1) + (1 - 1) + \cdots ,$$

then the sum appears clearly to be zero. However, the series can also be grouped in this way,

$$1 + (1 - 1) + (1 - 1) + (1 - 1) + (1 - 1) + (1 - 1) + \cdots ,$$

when the sum appears, just as clearly, to be 1. At this point a classically-minded mathematician might dismiss the series as improper and pathological, and indeed this was just the attitude of some mathematicians of the eighteenth and nineteenth centuries. Others, however, were more optimistic and more willing to experiment with their arguments to see where they led.

The great philosopher and mathematician Leibniz argued that since 0 and 1 are equally likely as the sum, depending on how many terms of the series you added, the actual sum of the series was $\frac{1}{2}$. This is not as absurd as it might seem. It fits other arguments. For example, call the sum of the series S, then:

$$S = 1 - (1 - 1 + 1 - 1 + 1 - 1 + 1 - 1 + 1 - \cdots) = 1 - S,$$

from which it follows that $S = \frac{1}{2}$. If that seems just another trick, here is an algebraic argument: consider the function $1/(1 + x)$ and its series expansion, which can be calculated using ordinary long division to divide 1 by $1 + x$:

$$1/(1 + x) = 1 - x + x^2 - x^3 + x^4 - \cdots .$$

This expansion is quite simply correct as long as x lies between -1 and 1. What happens when $x = 1$? The textbooks, with a 'classical' sense of what is proper, say that the series then does not have a meaning, but the 'romantic' mathematician will notice that the function itself has the value $\frac{1}{2}$ when $x = 1$! Who is right? Well, there is some right on both sides. The meaning of such series depends on the circumstances in which they are used, and how they are handled.

Mathematicians of the seventeenth and eighteenth centuries tended to use series with a gay abandon as if they were just like ordinary numbers and could be added, subtracted, multiplied and divided as you chose. As a result, they sometimes got into trouble, which was only sorted out when the ideas of the 'sum' of a series and what it means for a series to 'converge to a sum' were more deeply understood.

In other words, they misunderstood series rather as they misunderstood the idea of a polyhedron; in each case, the knots were untangled by deeper insight. Let us look at one example of these difficulties, and their resolution, in the theme of *divergent series*, about which still, in the early years of this century, 'there hung an aroma of paradox and audacity' (Hardy, in the Introduction to *Divergent Series*).

Divergent series include series such as

$$1 + 2 + 4 + 8 + 16 + 32 + \cdots$$

whose sums tend to infinity as you add more and more terms, so they are like the series $1 - 1 + 1 - \ldots$ but more so. Euler at one time thought that the sum of this particular series was -1. This might seem bizarre, but the fact was that quite often an argument that depended on the use of divergent series did actually 'work': it produced, in the end, the right answer. How could this be so if divergent series were as meaningless as some claimed?

Abel (1802–29), who proved that the general equation of the fifth degree cannot be solved using a finite number of arithmetical operations (including extraction of roots) only, wrote just before he died, 'Divergent series are the invention of the devil, and it is shameful to base on them any demonstration whatsoever.' The English mathematician Augustus de Morgan, writing in the *Penny Cyclopaedia* a few years later, was not so dismissive, but rather compared the mysteries of divergent series to the one-time mysteries of complex numbers, which by this time were well understood. Noting that, as we remarked, the use of divergent series did often produce the correct results, he suggested:

> If we then compare the position in which we stand with respect to divergent series, with that in which we stood a few years ago with respect to impossible quantities [that is, complex numbers], we shall find a perfect similarity ... It became notorious that such use [of complex numbers] generally led to true results, with now and then an apparent exception ... But at last came the complete explanation of the impossible quantity, showing that all the difficulty had arisen from too great limitation of definitions ...

De Morgan was right, it was a matter of definition, of understanding sufficiently clearly what you were talking about, as indeed Euler, with his wonderful imagination and insight, had realized a century earlier. Once again we come back to the question, 'When do mathematicians know what they are talking about?' and we see that it is intimately linked in practice to mathematicians' willingness to delve into the mysterious and the apparently absurd, in search of the simplicity and clarity and illumination that they so much enjoy.

Solutions to problems, Chapter 11

11A: *Langley's adventitious angles*

The original triangle ABC is isosceles, and since $A = 20°$, its angles at B and C are $80°$. Since we have constructed $E'\widehat{B}C$ to be $20°$, and $C = 80°$, we have $C\widehat{E'}B = 80°$ also, and so triangle $BE'C$ is isosceles.

Since $\widehat{ABC} = 80°$ and $\widehat{BCE} = 50°$, it follows that $\widehat{CEB}$ is also $50°$ and triangle BCE is isosceles.

Also, $\widehat{DBE'} = 60° - 20° = 40°$ and $\widehat{BE'D} = 180° - 80° = 100°$, so that $E'\widehat{D}B = 40°$, and so triangle $E'DB$ is also isosceles.

Therefore $BE = BC = BE'$ and, since $\widehat{EBE'} = 80° - 20° = 60°$, triangle BEE' is equilateral.

Therefore $EE' = BE' = DE'$, and so triangle $EE'D$ is isosceles, and $\widehat{EE'D} = 180° - 80° - 60° = 40°$ and $E\widehat{D}E = \frac{1}{2} \times 140° = 70°$. Finally, $E'\widehat{D}B = E'\widehat{B}D = 40°$, so $\widehat{BDE} = 70° - 40° = 30°$.

11B: *Prince Rupert's cube, and the puzzle of the divided square*

Prince Rupert's problem has two virtues, the simplicity and naturalness of the question, and the amazing answer: the largest square hole which can be cut in a given cube has an edge which is marginally under $\frac{3}{4}\sqrt{2}$ times the edge length of the original cube. Since $\frac{3}{4}\sqrt{2}$ is approximately 1.060660, it follows that you can pass a cube slightly larger than the original cube. Figure 11.20 shows how it is done: in this figure, the square hole cuts the top face along the lines *EFGH*, the bottom face along *ABCD* and the vertical edges at X and Y, as indicated by the dotted lines.

The three-quarters of a square is divided as in Fig 11.21. A simple puzzle, but it has fooled thousands of solvers over the years. Because the L shape is divided into four identical miniature versions of itself, this puzzle is related to other problems about *reptiles*, as they are called, such as the trapezium shown in Fig. 11.22, which also dissects into quarter copies of itself.

Fig. 11.20

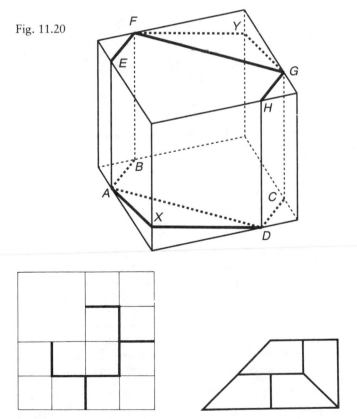

Fig. 11.21   Fig. 11.22

11C: *Julius Caesar's breath*

Very high! This surprising conclusion depends on a comparison between the number of molecules in one human breath, roughly estimated, and the number of breaths, measured by volume, which would make up the atmosphere. The former number is much larger than the latter, and so on average you can expect to have many molecules of Caesar's breath in each single volume of breath, even allowing for some of Caesar's breath having been chemically combined and removed from the atmosphere entirely.

11D: *The handcuff*

Figure 11.23 shows how.

Fig. 11.23

11:E *How many darts?*

This problem is the same in principle as the problem of the birth dates; but, because more people are used to playing darts than are accustomed to comparing birth dates, the result perhaps seems less surprising. The position of the first dart can be in any of the 20 sectors. The probability that the second dart will *not* strike the same sector is $\frac{19}{20}$. The probability that the third dart will *not* hit either of the first two is $\frac{18}{20}$. For the fourth dart, the probability is $\frac{17}{20}$, and so on. The chance that some pair of the six darts will stick in the same sector is 1 minus the probability that none of them do, so it is

$$1 - 19 \times 18 \times 17 \times 16 \times 15/20^5 = 1 - 0.436 = 0.564,$$

So, with just six throws, the chances of two darts hitting the same sector is greater than 50–50.

11F: *Smith and Watson's scores*

Consider the following sample figures for Smith's and Watson's performances.

	First half of the year		Second half of the year		Whole year	
	Watson	Smith	Watson	Smith	Watson	Smith
Runs	252	84	84	252	336	336
Times out	4	1	3	7	7	8
Average	63	84	28	36	48	42

Note the uneven distribution of Smith's scores and the fact that, if his scores for the first and second halves of the year had been switched, then Watson would have led in the first half of the year, Smith in the second, and the result would not have seemed at all paradoxical.

(This puzzle is no. 370 in the *Penguin Book of Curious and Interesting Puzzles*. The above explanation is adapted from Chapter 4 of Hugh ApSimon's *Mathematical Byeways in Ayling, Beeling and Ceiling* where there is a full discussion of the paradox.)

11G: *Two paradoxes of more than and less than*

The paradox appears because the everyday concept of more-than–less-than is being used to refer to ordinary sets of numbers, and infinite sets of numbers: but infinite sets and finite sets have different properties and behave in different ways. If you take any *finite* sequence of many consecutive integers then it is true that there will be more non-square numbers than square numbers, and the square numbers will only be a small proportion of the whole.

However, given any *infinite* set, it is *not* possible to count them one by one, as you can count a finite set, but it is possible to match the whole set to a part of itself, one for one. For example, here is the infinite set of integers, matched one for one against the even numbers:

1 2 3 4 5 6 7 8 9 10 11 12 13 14 ···

2 4 6 8 10 12 14 16 18 20 22 24 26 28 ···

Arnauld's paradox also depends on a failure to realize that more-than–less-than does not apply in the same way to positive and negative numbers together, as it does to positive numbers alone. If we stick to the positive numbers, then it is always true that if X is 'greater than' Y by subtraction, then X/Y is greater than 1. For example, $13 > 4$ and $13/4 > 1$. When negative numbers are introduced, this is no longer the case. Thus $13 > -4$ but $13/-4 = -3\frac{1}{4} < 1$. So we have to change our conception of what greater-than and less-than mean. If Arnauld had realized that this is often the case in mathematics, that new discoveries mean changing our old ideas, he would not have found his 'paradox' so puzzling.

11H: *Simplicity in calculation*

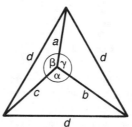

Fig. 11.24

Using Heron's formula for the area of a triangle is very complicated, because it involves squaring twice to get rid of the square roots. Adding the single construction line is more plausible, but the simplest method of the three is the third. Figure 11.24 shows again the right-hand diagram of Fig. 11.13. By the cosine formula applied to each of the small triangles,

$$d^2 = b^2 + c^2 - 2bc \cos \alpha, \qquad d^2 = c^2 + a^2 - 2ca \cos \beta,$$

$$d^2 = a^2 + b^2 - 2ab \cos \gamma.$$

We also know that $\alpha + \beta + \gamma = 360°$, and so $\cos(\alpha + \beta + \gamma) = \cos 360° = 1$. We therefore decide to use a well-known identity which connects $\cos \alpha$, $\cos \beta$ and $\cos \gamma$:

$$\cos^2 \alpha + \cos^2 \beta + \cos^2 \gamma = 1 + 2 \cos \alpha \cos \beta \cos \gamma.$$

Next, we substitute, like this:

$$\left(\frac{b^2 + c^2 - d^2}{2bc} \right)^2 + \left(\frac{c^2 + a^2 - d^2}{2ca} \right)^2 + \left(\frac{a^2 + b^2 - d^2}{2ab} \right)^2$$

$$= 1 + 2\left(\frac{b^2 + c^2 - d^2}{2bc} \right)\left(\frac{c^2 + a^2 - d^2}{2ca} \right)\left(\frac{a^2 + b^2 - d^2}{2ab} \right).$$

... and the rest is shear slog, illustrating the tedium of technique as well as its power. Hard work will now entirely substitute for insight and imagination! The only consolation as the mathematician ploughs through such algebraic manipulation is that the symmetry of the expressions makes it easier to spot mistakes in calculation, so the chance of having to do it all again is minimized. Leaping ahead, across a couple of pages of calculation, the final result is the equation shown in the text.

11I: *Roughly root 2*

If one fraction in the sequence is a/b then the next is $(a + 2b)/(a + b)$, and the sequence continues

$$3/2 \quad 7/5 \quad 17/12 \quad 41/29 \quad 99/70 \quad 239/169 \quad \cdots$$

By algebra, if

$$a^2 = 2b^2 + 1,$$

then $2a^2 + 4ab + 2b^2 = a^2 + 4ab + 4b^2 + 1$, that is

$$(a + 2b)^2 = 2(a + b)^2 - 1.$$

Notice that the adjustment is alternately 1 and -1; this means that the approximations are alternately greater than and less than $\sqrt{2}$.

11J: *Prime factorizations*

There is a connection between the two numbers, because

$$10^{12} - 1 = (10^6)^2 - 1 = (10^6 - 1)(10^6 + 1).$$

It is not difficult to discover with the aid of a calculator that

$$10^6 - 1 = 999999 = 3 \cdot 3 \cdot 3 \cdot 7 \cdot 11 \cdot 13 \cdot 37.$$

It is rather harder to discover that $1000001 = 101 \times 9901$, so that

$$M_{12} = 10^{12} - 1 = 999999999999$$
$$= 3 \cdot 3 \cdot 3 \cdot 7 \cdot 11 \cdot 13 \cdot 37 \cdot 101 \cdot 9901.$$

11K: *How do they rate?*

Table 11.2 shows the ratings given:

Table 11.2

	A	B	C	D	E	F	G	H	I	J
The teachers' scores	7	6	8	0	0	5	0	0	8	2
	6	7	8	7	4	3	6	9	2	7
	5	7	8	3	6	7	1	2	5	8
	9	0	4	0	8	0	0	6	0	8
	8	8	8	8	9	9	8	6	8	8
	6	9	7	9	9	5	2	6	2	7
	6	3	2	5	1	8	4	6	8	10
The teachers' averages	6.7	5.7	6.4	4.6	5.3	5.3	3.0	5.0	4.7	7.1
The professionals' averages	–	–	–	7.5	7.0	7.0	6.7	6.2	4.7	4.7

Chapter 12

A miniature world and
a long journey

It is difficult to give an idea of the vast scope of modern mathematics ... I have in mind an expanse swarming with beautiful details, not the uniform expanse of a bare plain, but a region of beautiful country, first seen from a distance, but worthy of being surveyed from one end to the other and studied even in its smallest details, its valleys, streams, rocks, woods and flowers.

Arthur Cayley (1821–95)

The centenary of Ramanujan's birth was celebrated by a conference at which the celebrated physicist Freeman Dyson contributed a paper titled 'A Walk Through Ramanujan's Garden', which he concluded thus: 'Whenever I am angry or depressed, I pull down [Ramanujan's] collected papers from the shelf and take a quiet stroll in Ramanujan's garden. I recommend this therapy to all of you who suffer from headaches or jangled nerves. They also are full of beautiful ideas which may help you to do more interesting mathematics.'

No mathematical problem or theme is an island, separated from the main body of mathematics. It is true that certain corners of mathematics, such as the snowflake curve and other 'pathological' curves, once seemed to be isolated and alone, yet even in their isolation they were studied, questions were asked about them, novel ideas created, and their tiny corner of the mathematical landscape cultivated.

Any problem, from the simplest to the most magnificent, is a door into this mathematical world. Like Alice stepping through the Looking Glass or disappearing down the rabbit hole, we find ourselves on the other side in a miniature world of strange objects, fascinating features, and attractive ideas, which we are invited to explore. One idea leads to another, and soon we are likely to find ourselves travelling away from our point of entry, perhaps into realms which we never even knew existed. We have already glanced at a few of these miniature worlds in this book, and made a few very short trips, for example from the problem of the point in the equilateral triangle to Napoleon's theorem, and then to other more general theorems (pp. 178–84).

In this chapter, we shall first look at one small area in the mathematical landscape, then survey a rather larger region, and finally set out on a journey that will take us from the Greek mathematicians to quantum theory.

Centroids

Figure 12.1 is the same as that for Problem 9E. This problem asked for the *centroid* (the mathematical term for the centre of gravity) of the L shape to be found by the use of a straight edge only, that is, with an unmarked ruler. We are going to stick with this problem, without moving very far away, and see what other possibilities it suggests. The solution to that problem was to split it in two ways into the sum of two rectangles. Each time, the centroid of the L shape will lie on the line joining the separate centroids as in Fig. 12.2. Given this method, it is natural to wonder if it can be applied to more

Fig. 12.1

Fig. 12.2

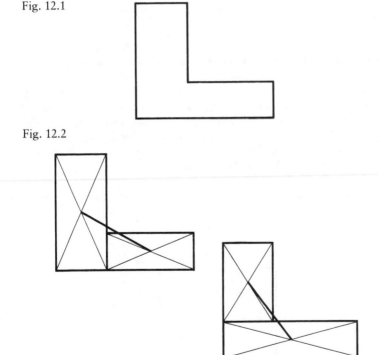

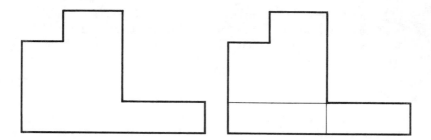

Fig. 12.3

complicated shapes, such as in Fig. 12.3. Yes, it can, by dividing the figure by either of the thin lines, as on the right, into an L shape and a rectangle, in two different ways, and then proceeding as before. (You will notice that there are also two other ways to split the figure into an L shape and a rectangle.)

The L shape can be regarded in a different way, as a rectangle with a rectangle cut from one corner. In other words, we have been able to find the centroid of a rectangular shape from which a rectangle has been *subtracted*. This suggests that, by the use of a straight edge alone, we can find the centre of gravity of any shape which can be made by adding or subtracting rectangles from rectangles, including such complicated shapes as in Fig. 12.4. Indeed we can, though the number of lines to be drawn is now becoming very large, and the accumulation of errors will *in practice* be considerable. However, we do not need to worry about these practical difficulties while we are pursuing a theoretical train of thought.

Fig. 12.4

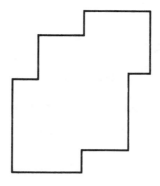

Problem 12A   *The missing rectangle*

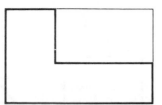

Fig. 12.5

What is the relationship between the centroid of the L shape in Fig. 12.5, as found by the text method, and the centroids of the whole rectangle and the rectangle 'missing' from one corner?

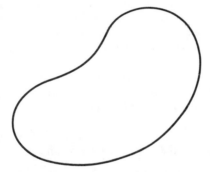

Fig. 12.6

The fact that we can find the centroid of very complicated shapes suggests that we might *approximate* a curved region (such as in Fig. 12.6) by a set of rectangles and so construct its centroid approximately, by straight edge alone. This requires, however, something we have not yet considered. How do we construct a set of rectangles? The least equipment that we require is set-square, to provide us with right angles, and a parallel rule (as in Fig. 12.7) so that we can draw

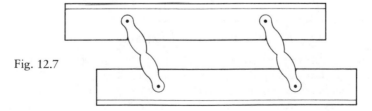

Fig. 12.7

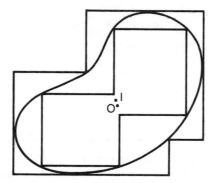

Fig. 12.8

opposite sides parallel. Figure 12.8 shows the same shape approximated by a set of interior rectangles, and an outer boundary which also encloses a set of rectangles. The centroids of the two different sets are marked as O for outer and I for inner. The fact that they are not too far apart suggests that they are fair approximations to the actual centroid.

The original L shape can be thought of as a rectangle with a rectangle missing. What would happen if a larger part of the original shape were missing? What would happen if it in fact consisted of two separate parts?

The rectangles shown in Fig. 12.9 are quite separate, though they are related by the fact that their sides are parallel. Consequently, when

Fig. 12.9

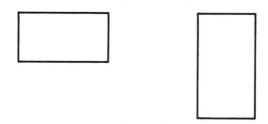

Problem 12B *Rectangles or parallelograms?*

A parallel rule can only be relied upon to draw parallelograms. For our purpose of finding the centroid of a region, by dividing it into small simple pieces, why is a parallel rule alone sufficient?

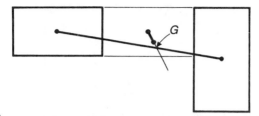

Fig. 12.10

we add the thin lines as in Fig. 12.10, we are constructing another *rectangle*, which like the original missing corner can be thought of as having been subtracted from an original figure which contained all three rectangles.

Can we find the centroid of the pair of rectangles? Yes, by finding the centroid of the whole figure, and the centroid of the missing rectangle. The centroid of the pair of rectangles will lie on the line joining those two points; but it will also lie on the line joining the centres of the two rectangles themselves, and so it is the intersection of these two lines, at G. This in turn means that we can find the centroid of any set of suitably placed rectangles, or of shapes composed of a set of rectangles.

So far we have only considered plane figures. What about three-dimensional figures formed by sets of cuboids? In other words, solid bodies whose faces are all planes in three directions at right angles to each other. (Or just in any three given directions, given that parallelograms are just as effective as rectangles.) Figure 12.11 is an analogue

Fig. 12.11

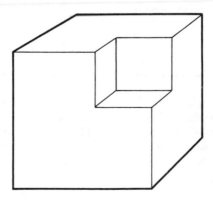

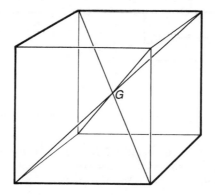

Fig. 12.12

of our original figure: a solid cuboid with a cuboid cut out of one corner. Can we construct its centroid with a straight edge? Yes, bearing in mind that we need imagination, or a physical model, to actually draw lines in three dimensions. The centroid of a cuboid will be the point where all *three* of its space diagonals intersect; see Fig. 12.12.

One solution is firstly to find the centroid of the L shape in the top layer (see Fig. 12.14), and the centroid of the bottom layer which is just a rectangular box, and then to join these two centroids with a

Problem 12C   *Relative areas*

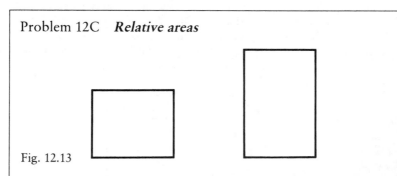

Fig. 12.13

The relative weights of the two rectangles in Fig. 12.13, which in this case just means their relative areas, will determine where on the line joining their centres the joint centroid lies. Without therefore actually measuring the sides of either rectangle, how can their relative areas be represented as the ratio of two line segments?

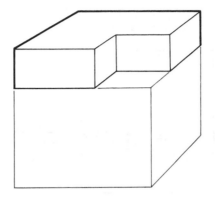

Fig. 12.14

line. The centroid of the whole will lie on this line. Then we could slice the shape parallel to the right-hand face, through the left-hand face of the 'missing' corner, and so split it into a different L shape (on the right) and a cuboid. Join their centroids and the overall centroid will be where this line meets the first line.

This exploration has been very localized in the mathematical landscape, as it were, because we have resolutely stuck to the use of a straight edge only. Even with this very severe limitation, there are connections to other themes: to mention just two: Georg Mohr, a mathematician with no other claim to fame, published the surprising discovery in 1672 that every construction that can be made with ruler and compasses can effectively also be made with compasses alone. (A complete straight line cannot be drawn with compasses, of course, but any point which can be constructed on a straight line, can be constructed with compasses as the point of intersection of two arcs.)

This fact was rediscovered in 1797 by the better-known Lorenzo Mascheroni (1750–1800) and such compass-only constructions are called Mascheroni constructions after him. In 1805, the little-known geometer Servais published the result that *any* construction in the plane that can be performed with a straight edge and a pair of compasses can be performed with a straight edge alone, provided that one circle and its centre is given. (This result was rediscovered in 1822 by the famous mathematician Poncelet (1788–1867), to whom the result is usually attributed.) However, instead of following these leads, we shall turn aside to an entirely different theme.

Problem 12D   *Overlapping rectangles*

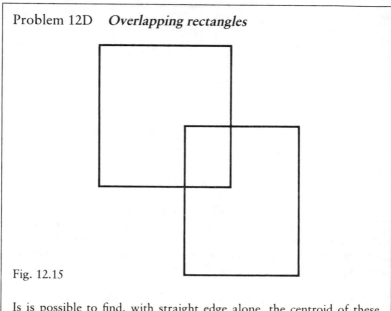

Fig. 12.15

Is is possible to find, with straight edge alone, the centroid of these two overlapping rectangles shown in Fig. 12.15, taken together?

Extremes, and reflection

According to one legend, when Aeneas on his wanderings after the fall of Troy arrived on the shores of North Africa, 'they reached the place where you will now behold mighty walls and the rising towers of the new town of Carthage; and they bought a plot of ground named Byrsa . . . for they were to have as much as they could enclose with a bull's hide.'

According to another version of the story, it was Queen Dido of Carthage herself who marked out the citadel of Carthage by the same means. Whatever the truth, it led to the famous mathematical problem illustrated in Fig. 12.16. The straight line represents the seashore. The hide has been cut, naturally, into thin strips which have been tied to

Fig. 12.16

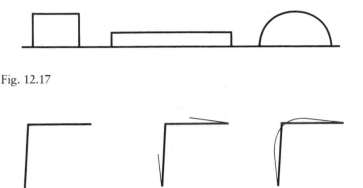

Fig. 12.17

Fig. 12.18

form a long rope. The problem is reduced to surrounding the largest area against the seashore with a given length of rope. Which of these arrangements of the rope shown in Fig. 12.17 will enclose the greatest area? It is not difficult to conclude that sharp angles in the rope will not be efficient: given an angle as in the left-hand diagram of Fig. 12.18, we can increase the area slightly, and make a smooth curve with the same perimeter but greater area. But what smooth curve is best?

The clue lies in the discovery of Pappus (*c.* AD 300) that, of all *closed curves* with a given perimeter, the circle has the largest area. The elegant trick which goes with this clue is to reflect the curve which we suppose to be a maximum, in the shore line, to obtain just a closed curve (Fig. 12.19). The perimeter of this closed curve is fixed, at double the length of the rope. Its area, and therefore half of its area, will be a maximum when it is a circle, and therefore when one half of it is a semicircle.

Fig. 12.19

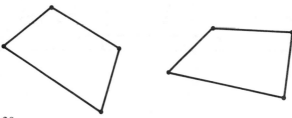

Fig. 12.20

The ideas in this solution can be used to solve a variety of related problems. Let's look at some examples. Figure 12.20 shows two positions of a quadrilateral which is hinged at its vertices. As the sides are moved, its area changes. When is it a maximum?

We can adjust the sides so that the vertices lie on a circle (Fig. 12.21). Imagine that this has been done, and that the portions of the circle outside the quadrilateral are solid and attached to the outside of the quadrilateral, but still allowing the quadrilateral to hinge as

Fig. 12.21

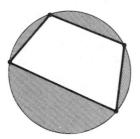

Problem 12E   *Rabbi Ben Moses' box*

The Greeks also supposed that, of all solid shapes, the sphere encloses the greatest volume for a given surface area. By analogy, and symmetry, it is natural to expect that, of all rectangular boxes, the cube encloses the greatest volume for its surface area.

Given this assumption, and the area of *reflection*, how is it possible to deduce Rabbi Solomon Ben Moses' result (Problem 7D) that the rectangular box with a square base and an open top that has maximum volume for its surface area has a height equal to one half of the edge of the base?

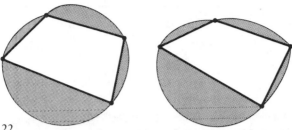

Fig. 12.22

before. Then, as the quadrilateral changes shape (see Fig. 12.22), the perimeter of the figure does not change, and the area of the attached segments does not change, but the area of quadrilateral does. The area of the quadrilateral will be greatest therefore when the area of the whole figure is a maximum, which will be when it is a circle. Therefore the area of a quadrilateral is a maximum when it is *cyclic*. This conclusion can be generalized: any polygon whose sides are given in order but whose angles are variable has its maximum area when it is inscribed in a circle.

Figure 12.23 shows a plan of a screen standing in the corner of a room. The screen has two halves, hinged where they join. How should the screen be placed to maximize the area enclosed? This time, the trick is to reflect the screen twice, in each wall, and then once more to create an octagon, as in Fig. 12.24. The area of the octagon is

Fig. 12.23

Fig. 12.24

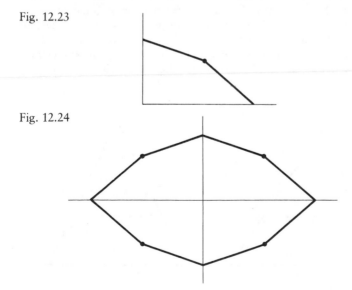

Problem 12F   *The maximum isosceles triangle*

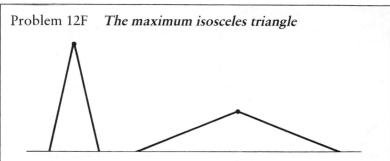

Fig. 12.25

An isosceles triangle has two equal sides of given length, but the angle between them is variable (Fig. 12.25). How does the reflection principle indicate the angle for which the area will be a maximum?

---

four times the area enclosed by the screen, and they will be maximized together. The octagon will have maximum area when it is a regular octagon, inscribed in a circle. Therefore the screen encloses the maximum area when the angles between the two halves and between the screen and the walls are 135° and 67½° respectively.

Notice that we have used a reflection trick here, but it is not the same as the reflection principle used to solve problems about random walks (p. 213). The transformation of reflection is often useful in mathematics. We shall see yet another use of a third 'reflection principle' later in this chapter.

Figure 12.26 illustrates the problem of finding the largest rectangle which can be fitted into a given triangle. What shape should the

Fig. 12.26   Fig. 12.27

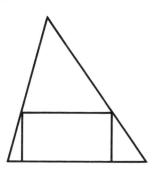
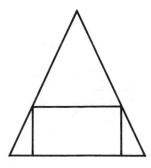

rectangle be? To solve this problem, it helps to use the concept of a *shear*. The idea is that, if the triangle is *sheared* sideways, then the area of the triangle will not change, and the area of a rectangle resting on the base of the triangle will not change, provided that it moves with the triangle. Figure 12.27 shows the result of a shear, and the second rectangle has the same area as the first, although its base has had to slide a short distance along the base of the triangle.

Using this idea, we can replace our original triangle with this isosceles triangle, and then ask the same question. However, we can now go further in our quest for a *simple* problem by squashing the triangle vertically. (An equivalent argument would apply if we had started with a flatter triangle and stretched it vertically instead of squashing it.) Squashing the triangle will change its area, and it will also change the area of any inscribed rectangle on its base, *but it will change them in the same proportion.* Thus, if the original rectangle is the largest that can be inscribed in the triangle, then the squashed rectangle will have the same property in relation to the squashed triangle (and conversely). So, with this idea in mind, we squash the isosceles triangle until it has a right angle at the top and is in fact half of a square (Fig. 12.28). Finally we *reflect* the half square in its hypotenuse, and ask ourselves what shape of rectangle inscribed in this square so that it is reflected in a diagonal of the square will have maximum area?

The answer depends on the fact that, *whatever* rectangle is inscribed under those conditions, the sum of two adjacent sides will be constant.

Fig. 12.28

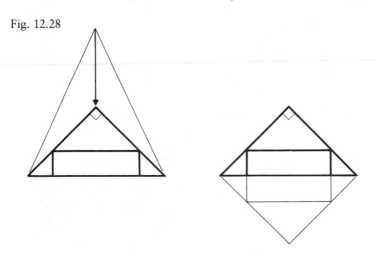

(This would not have been true of the figure obtained by reflecting the original triangle in its base line, before we had sheared and squashed it.) We know already that, if the sum of two numbers is constant, then their product is a maximum when they are equal. Therefore the rectangle has maximum area when it is a square, and one half of it is half a square with half the height and half the width of the triangle.

Reversing the squashing and shearing, we conclude that the maximum rectangle inscribable in a triangle is one half its height and one half its width (Fig. 12.29). In fact, it can be obtained by joining the midpoints of the sides and dropping perpendiculars to the base. (Since each side of the triangle can be regarded as its base, there are actually three such rectangles, each having half the area of the original triangle).

We can wander a little further along this mathematical route. What is the maximum rectangle that can be inscribed in a circle? Suppose that this rectangle has sides $2x$ and $2y$ (Fig. 12.30), and let the radius of the circle be r; then we know by Pythagoras that $x^2 + y^2 = r^2$. The area of the rectangle is $4x^2y^2$, and so we are looking for the maximum value of a product x^2y^2 when we know that the sum of x^2 and y^2 is constant. As usual, the answer is that the product is a maximum when the two numbers are equal: $x^2 = y^2$, and so $x = y$, and the rectangle is a square. No surprise there!

Fig. 12.29

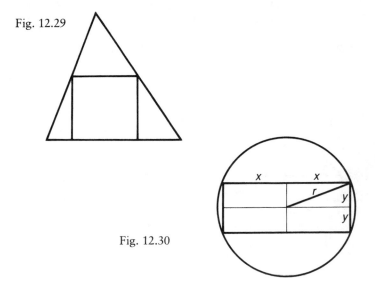

Fig. 12.30

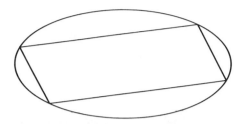

Fig. 12.31

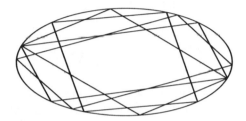

Fig. 12.32

What happens, however, if we now either *squash* or *shear* the circle? Either transformation will turn the circle into an ellipse, and will also transform parallel lines into parallel lines, so that a rectangle becomes a parallelogram (Fig. 12.31). Conversely, we can take any ellipse, and an inscribed parallelogram, and shear it or squash it back into a circle, turning the parallelogram into a rectangle. What does this tell us about the maximum parallelogram that can be inscribed in an ellipse? The circle can have its inscribed square at any angle at all. When squashed or sheared into an ellipse, all these squares, which are equal in area, will become different-shaped parallelograms, all of maximum area. We can therefore conclude (see Fig. 12.32) that there is not one single maximum parallelogram which can be inscribed in an ellipse, but a whole family of them, with different angles, but all with maximum area. Now that is a surprise!

From Heron to Feynman

A deservedly popular puzzle asks for the shortest journey that Jack can take from *A* to *B*, visiting the river bank on the way (see Fig. 12.33). The solution is simple and elegant: reflect *B* in the line of the

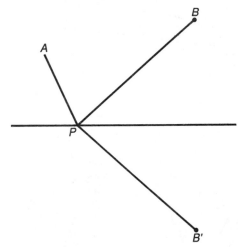

Fig. 12.33

river bank, to B'. Then $AP + PB = AP + PB'$ and $AP + PB$ will be a minimum when $AP + PB'$ is a minimum and therefore when APB' is a straight line, and the angles between AP and PB and the river bank are equal.

This is a problem about *extremes*, the greatest, least, longest, fastest, shortest, hottest ... That extremes are attractive and intriguing is witnessed by history, by people's delight in the fabulous and legendary, and more recently by the success of the *Guinness Book of Records*, now available in Russian and Chinese versions, as well as most European languages and five Indian languages.

The very idea of extremes is challenging, though it must be admitted that there is an element of sex and age bias here: the *Guinness Book of Records* sells mostly to teenage boys. Fortunately, extremal problems also often display great elegance and simplicity, both in their expression and in their solutions. Just as importantly, they are an essential feature of all the sciences.

Two thousand years ago, Heron of Alexandria (*c.* AD 50) discussed a problem analogous to Jack's trip to the river and back, in these words: 'Practically all who have written ... of optics have been in doubt as to why rays proceeding from our eyes are reflected by mirrors and why the reflections are at equal angles.' Heron refers to 'proceeding from our eyes' because that is what the Greeks thought that rays of light did. They were wrong, but that makes no difference to Heron's solution to the problem. He argued that light necessarily travels in straight lines, and then noted that a straight line is the shortest distance between two points: he then reversed this observation to conclude that rays of light travel in such a way that they go the minimum distance possible, even when being reflected from a mirror. He then proved that, on this assumption, the angles of incidence and reflection are equal, using the same argument that we have used to solve Jack's puzzle.

Jack's puzzle by itself might be nothing more than a little mathematical teaser, amusing, but otherwise of no value. Heron's problem, however, which is its exact analogue, is something else entirely: it would have been important merely as the first application of mathematics to optics. In fact it is much deeper than that.

We have seen that the Greeks knew that circle was the curve of least perimeter enclosing a given area. Aristotle, long before Heron, argued from this fact that the planets naturally moved in circles, because 'of lines which return upon themselves the line which bounds the circle is the shortest'. Aristotle's argument here is false, but, as

---

Problem 12G   *The advantageous race*

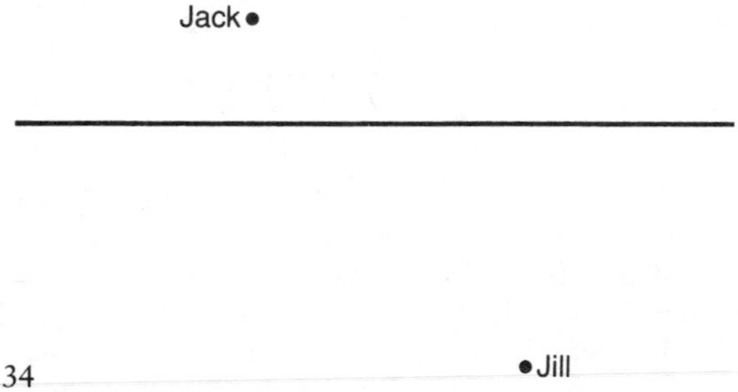

Jack●

Fig. 12.34   ●Jill

Jack and Jill are standing on opposite sides of the path, marked by the straight line, when they agree to have a race to a point on the path (Fig. 12.34). Jack is already nearer the path than Jill, so he has an initial advantage. Not content, however, merely to win, he decides that they will race to the point on the path that makes the difference between the distance that Jill has to run and the distance that he has to run as large as possible. Which point does Jack suggest as their goal? Jill, of course, wishes to run a fair race, so she suggests that they run to a point on the path which is equidistant from each of them. Which point is this?

Problem 12H   *Jack's puzzle, revised*

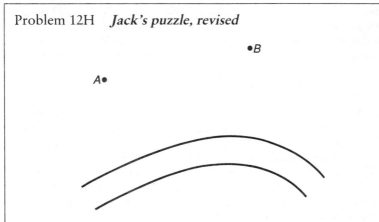

Fig. 12.35

In Fig 12.35 Jack's problem is the same in that he must start from A, visit the river to collect water and go to B; however, the river is not straight now but bends. What is the connection between the solution to this problem, and the properties of the ellipse, as presented in Chapter 7?

has happened so often in the history of science, this false argument contains the germ of a sound idea – that, as Isaac Newton put it, quoting a familiar opinion, 'the philosophers say that Nature does nothing in vain, and more is in vain when less will serve; for Nature is pleased with simplicity . . .' Gottfried Leibniz (1646–1716), Newton's contemporary and rival and co-inventor of the calculus, was even keener on the same idea, arguing that our world is the best of all possible worlds, exhibiting 'the greatest simplicity in its premises and the greatest wealth in its phenomena.'

Heron was following the same idea as Aristotle when he argued that the path of a ray of light is as short as possible. However, he made a mistake in supposing that it was the *distance travelled* that was minimized. If that were indeed so, then a ray of light passing from air into water would not deviate from its original straight path. Yet it does deviate, as we saw in Chapter 7, in such a way that the *time taken*, not the distance moved, is a minimum. This 'principle of least time' was first proposed by Fermat, who used his principle, and some other novel ideas, to deduce Snell's law.

Our story so far sees, over a period of two thousand years, Aristotle's intuition, based on entirely false reasoning – so typical of the development of science – followed by Heron's correct reasoning in a situation in which 'least time' and 'least distance' are equivalent, followed by Fermat's principle which explains the refraction of light by liquids. Once again, this could have been the end of the journey, but it was not. In 1744, the French mathematician Maupertuis (1698–1759) published his *Principle of Least Action*. Once again, the subject of the 'principle' changed. Having previously been distance, and then time, it now became 'action' which Maupertuis defined, for a single particle, vaguely and unsatisfactorily, as the product of the mass, the velocity and the distance. To support his 'principle', he produced detailed examples, which were of course all in his favour. Examples which might have undermined his principle were ignored.

Yet, once again, intuition proved, in the long run, to be stronger than any immediate practical objections. In the very same year, 1744, Euler published a mathematical theorem which expressed the same idea. Euler was also inspired by preconceptions about nature: 'Since all processes in nature obey certain maximum and minimum laws, there is no doubt that the curves that bodies describe under the influence of arbitrary forces also possess some maximum or minimum property.' Euler went on, much more like a modern scientist, to observe that while it was impossible to deduce what this *extremal* principle might be from philosophical principles, experiment provided the evidence from which the correct extremal principle might be inferred.

At last, the idea which started with an intuition on the part of the ancient Greeks, was on a scientific footing, presented not as a property of God or Nature, but as a hypothesis supported by experiment. It was further developed by Lagrange (1736–1813) and in the nineteenth century by Hamilton, whom we have already met in connection with the Icosian Game and Hamiltonian paths, and after whom 'Hamilton's principle' is named. In the twentieth century, Richard Feynman gave his name to the 'Feynman principle' in quantum mechanics, which is related to the original principle of least action. In between Hamilton and Feynman, the great physicist Helmholtz had expressed the view that the principle of least action offered a unifying principle not just for optics and mechanics but for the whole of physics.

How marvellous that an idea which appealed originally as much on aesthetic or metaphysical grounds, as on any practical success, should turn out to be of profound practical importance! As many scientists have observed, in wonder and puzzlement, the simplest and most beautiful ideas do seem to have the best chance of being true.

Having sketched this mathematical journey in a straight line from Aristotle to Feynman, let us now return to Fermat and explore another route, which also involves minimum distances. The following problem was posed by Fermat to Galileo's pupil Torricelli (1608–47) who invented the barometer: 'Find the point such that the sum of its distances from the vertices of a given triangle is a minimum.' The solution to this problem was anticipated in Chapter 7, Problem 1G, 'A shortest network', and is illustrated in Fig. 12.36. It is practical, because it represents in very simplified form the problem of minimizing the costs of travel, road transport, even road building. The solution, provided that each angle of the triangle is less than 120°, is to draw three roads, through the vertices of the triangle, which meet each other at 120°, as we noted for Problem 1G. But why is this? We can start by marking three points to represent the corners of the triangle (the edges are not needed), as in Fig. 12.37. We pick a point X and suppose that this is the point which makes $AX + BX + CX$ a minimum. Next we imagine that A is moved to a new position on the line AX, say A', and we wonder what point will make its sum of distances from A', B and C a minimum? Will it be the same point X, or will there be a new and different minimum point at, say, X'?

Fig. 12.36

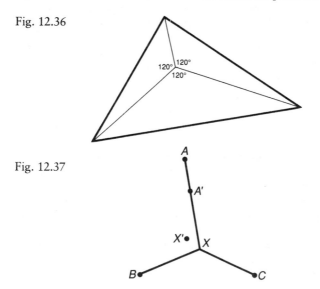

Fig. 12.37

Suppose that the new minimum is indeed the new point X'. Then we know from the original three points that

$$AX + BX + CX < AX' + BX' + CX',$$

and we know from the three points A', B and C that

$$A'X' + BX' + CX' < A'X + BX + CX.$$

Adding, and cancelling, we find that

$$AX + A'X' < AX' + A'X.$$

Substituting $AX = AA' + A'X$ and cancelling again, we conclude that

$$AA' + A'X' < AX'.$$

But this conclusion is absurd. Any one side of a triangle is less than the sum of the other two, so $AX' < AA' + A'X'$, and not the other way round. Therefore our original supposition, that the minimum point for A', B and C was *not* X is false: therefore the minimum point for A', B and C is the same as for A, B and C.

It follows that, however we replace A by A', or B by another point on BX, or replace C similarly, the minimum point X never changes. In particular, if we adjust the original A, B and C to three new positions that make an equilateral triangle, then X will not change. But, by symmetry, in an equilateral triangle it will be at the centre, and the three lines will meet at 120°. Therefore our original conclusion is proved, with, however, two provisos. We have assumed that there is just one minimum point – if there were more, for example if there were three minimum points, then they could be arranged symmetrically about the centre of the equilateral triangle. And we have assumed that the original point A was adjusted to a position A' *between* A and

---

Problem 12I *Weighted routes*

Suppose that three straight roads are planned to connect towns A, B and C, but with the condition that, while the amount of traffic between A and C equals the amount between A and B, the traffic between B and C is double that amount. As before, the layout should minimize the number of vehicle-miles (or vehicle-kilometres) travelled. Will it still be true that the angles which the three roads meet is independent of changes in the position of the three towns along the previously defined roads to the network centre?

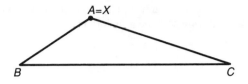

Fig. 12.38

X. If A were changed for a point beyond X, then the proof would no longer work. In fact, if the original triangle has a vertex with an angle equal to or greater than 120°, then the minimum point is at that vertex (Fig. 12.38).

This proof is simple and has the desirable property of depending only on the most basic inequality about distances, that one side of a triangle is shorter than the sum of the other two. It follows at once that the same conclusion can be drawn about the analogous problem on a sphere (such as the surface of the earth), because the same inequality is true on a sphere when 'straight lines' are defined to be arcs of great circles.

How can we construct the Fermat point of a given triangle? A simple construction is to draw equilateral triangles on the outside of each edge, and then draw their circumcircles, as in Fig. 12.39. These meet in a point, which is the Fermat point of the triangle. Joining it to the vertices, we can see why the three lines meet at 120°: the angle at which any two meet is the angle of a cyclic quadrilateral whose

Fig. 12.39

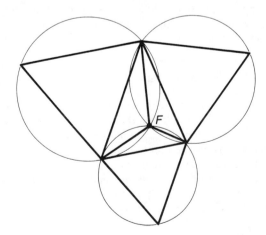

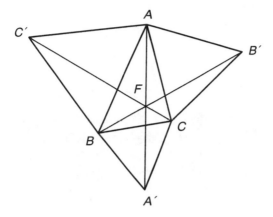

Fig. 12.40

opposite angle is 60°, and opposite angles of a cyclic quadrilateral sum to 180°. You will notice that Fig. 12.39 closely resembles the basic figure for Napoleon's theorem, and displays another property of that figure. Figure 12.40 illustrates a third property: AA', BB' and CC' concur at the Fermat point.

The same question was posed for more than three points by Steiner (1796–1867) after whom the solutions are called Steiner networks. What is the shortest road network for four points? The idea of constructing equilateral triangles on the outsides of the edges works here also, provided that the quadrilateral is sufficiently 'ordinary', as Figs 12.41 and 12.42 illustrate. In Fig. 12.41, the total length $CQ + DQ + QR + RB + RA$ equals PS (because of the conclusion of

Fig. 12.41

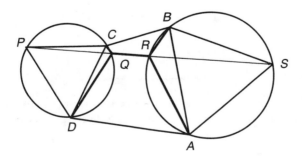

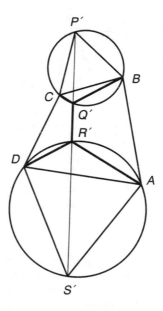

Fig. 12.42

Problem 4L). In Fig. 12.42 the total length of the network is $P'S'$. To find the absolute minimum for the original four points, we find PS and $P'S'$ and choose whichever is the shorter.

Problem 12J   *The Steiner network for a square*

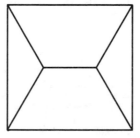

Fig. 12.43

Because of its symmetry, a square has two Steiner networks (Fig. 12.43 shows one of them), formed of new line segments meeting at 120°. What is the length?

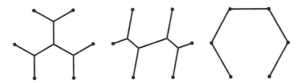

Fig. 12.44

What about larger Steiner networks? For a symmetrical arrange-ment, such as the vertices of a regular hexagon, it is not difficult to calculate exact solutions. Figure 12.44 shows three, each of a different pattern. Their total lengths, from left to right, measuring the distance between adjacent vertices as 1 unit, are $3\sqrt{3}$, $2\sqrt{7}$ and 5 units respectively. It is disappointing to notice that the elegant solutions left and middle are actually longer than the boringly obvious solution on the right!

Larger minimum networks can be constructed by direct calculation by computer if the number of points is not more than about 35. Minimum networks, or close approximations to a minimum, for even larger numbers of points can be found by using, surprisingly enough, soap solution! Soap bubbles form minimum surfaces because of their surface tension. Figure 12.45 shows the result of dipping a cubical framework into a soap solution. This is the smallest surface, in area, that will join all twelve edges of the cube. The small internal 'cube' actually has slightly bulging edges.

Fig. 12.45

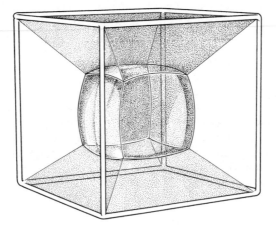

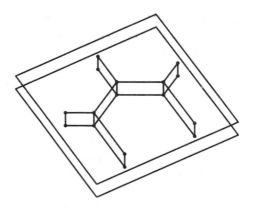

Fig. 12.46

To solve a minimum-route problem using soap solution, we take two transparent sheets and join them by pins, placed to represent the individual towns. The double sheet is then dipped into soap solution, and the resulting films will meet at 120° everywhere (Fig. 12.46). For any three pins, the soap solution will find the Fermat point. For four or more points it will find a solution which is a *local* minimum – meaning that a slight variation will not reduce the total length, but there is no guarantee that a completely different way of joining the pins will not produce a shorter length, as we found with four points.

Steiner networks are enjoyable in themselves, as well as being over-simplified mathematical *models* of actual situations. Real life, of course, is more complex. If roads are built to suit the road user, then every town might be joined to every other town. If one town is the capital city, then every other town might be joined to it, whether or not they are joined to each other. The Steiner network may suit the taxpayers who pay for the roads to be built, but may not suit the same taxpayers when they are driving a car and find that they cannot drive on a straight road between any pair of towns.

At this pause in our journey across the mathematical landscape, we could choose to investigate these real-life problems. Instead we shall turn down a different road, return to Heron's interest in bouncing light rays, and look at a problem posed by the eighteenth-century mathematician G. F. Fagnano. The problem is to inscribe in a given triangle, the triangle of *minimum perimeter*. Figure 12.47 shows a

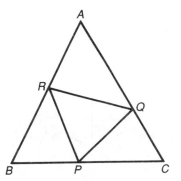

Fig. 12.47

plausible attempt at a solution. What can we say about it? If the whole inscribed triangle has minimum perimeter, then RPQ must be the shortest route from R to Q, via the edge BC, and so RP and PQ make equal angles with BC, as if they were the path of a light ray bouncing off a mirror. Similarly, the angles of incidence and reflection at Q are equal, and so are the similar angles at R. We can now go further with a little angle chasing. Mark the angles as in Fig. 12.48. Then $A + y + z$, $B + z + x$, $C + x + y$ and $A + B + C$ all sum to 180°. Adding the first three and subtracting the fourth, and then halving, we find that $x + y + z = 180°$ also, and so $A = x$, $B = y$ and $C = z$. As it happens, this is a property of the triangle formed by the feet of the altitudes (Fig. 12.49) and the converse is also true: so the solution to Fagnano's problem is the *pedal triangle* formed by the feet of the altitudes.

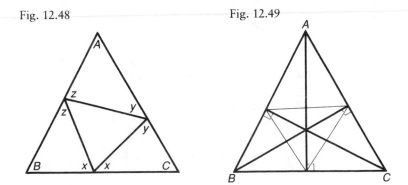

Fig. 12.48   Fig. 12.49

Problem 12K   *An altitude property*

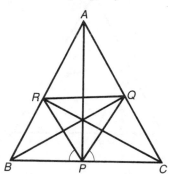

Fig. 12.50

The feet of the altitudes have been joined to form the pedal triangle shown in Fig. 12.50. Why does it follow from the properties of circles that the marked angles at *P* equal *A*?

It is not a great surprise that a similar solution exists for the quadrilateral of minimum perimeter inscribed in a given *cyclic* quadrilateral. The qualification 'cyclic' is essential. Many properties of the triangle can be neatly adapted to cyclic quadrilaterals, but not to quadrilaterals in general. Anyway, it is obviously useless to construct altitudes from the vertices of the quadrilateral to its sides; but what we do instead is find where the diagonals meet, and drop altitudes from this intersection to the edges. The feet of these altitudes are the vertices of the minimum-perimeter inscribed quadrilateral (Fig. 12.51).

Fig. 12.51

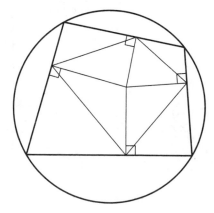

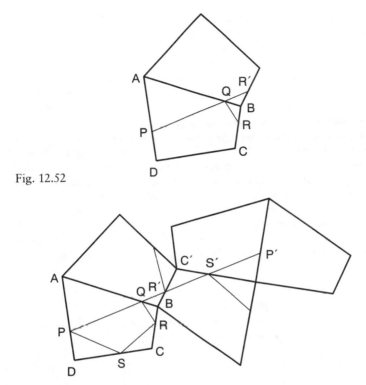

Fig. 12.52

Fig. 12.53

If an ordinary quadrilateral is not cyclic, then a minimum solution exists but cannot be found by this simple method. What we can do is to rely, yet again, on reflections, as in Fig. 12.52. We imagine the light ray starts at P and first bounces at Q. At this stage we reflect the whole quadrilateral about the side AB, so that the actual path of the light ray is QR but its reflected path is along the straight line PQR'. It next reflects at R, and we can show this as a continuation of the same straight line, by reflecting the quadrilateral again about BC'. Finally it reflects at S and returns to P, and we show this by reflecting the quadrilateral for the third time, as in Fig. 12.53. The total length of the path is the length of PP'.

In our journey we have oscillated between applications – very real problems about light reflecting from mirrors, or being refracted through water, or the behaviour of particles in dynamics – and pure problems such as Fermat's question to Torricelli, or Fagnano's problem, and back to applications, such as Steiner networks. This is typical of the development of mathematics, in which pure questions and questions about the real world are intimately related, each prompting and provoking the other. We shall conclude this chapter and end our journey by looking at a practical application of light rays bouncing around inside closed curves.

In physics, a *billiard* in two dimensions is a closed curve with a smooth boundary – the 'billiard table' – together with a point which repeatedly reflects around the inside of the curve, like a ray of light. This apparently simple model is very useful for considering problems in dynamics, optics, acoustics and other areas of science.

G. D. Birkhoff, who originated this theory, showed that, if the billiard table is convex, then there are paths with any number of bounces (greater than one) which repeat themselves. For two bounces it is sufficient to simply take the longest chord which can be drawn

---

Problem 12L   *A billiards puzzle*

Fig. 12.54

A billiard ball flies out of the bottom left hand corner, at 45° to each of the edges, and bounces round the table shown in Fig. 12.54. The table is perfectly smooth, and the cushions are perfectly elastic, so the ball never loses its initial velocity and every bounce is a perfect reflection, the angle of incidence equalling the angle of reflection. The table, as you can see, has no pockets along the sides, only one pocket in each corner. If the table is 9 units wide and 7 units deep, will the ball continue bouncing for ever or will it end up in one of the pockets? If the latter – which pocket?

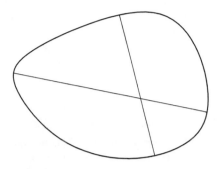

Fig. 12.55

inside the curve (Fig. 12.55). The longest chord is a *diameter* – that is, it cuts the curve at each end at right angles (because otherwise a slightly longer chord could be drawn slightly to one side) – and so a ray of light travelling along this diameter will be reflected back along the chord at each end. Similarly, the shortest chord that can be drawn is also a diameter, so it is another periodic path.

Paths with three or more bounces are harder to find (except in circles). Figure 12.56 shows two paths with five bounces each on an elliptical billiard table. Such paths are pretty, but they are also rather simple – too simple for the physicist. More important are paths which do *not* repeat, in particular, paths which have the property of passing arbitrarily close to any point inside the curve, and doing so, sooner or later, arbitrarily close to any given direction. This certainly will not happen on a rectangular billiards table, because the rays of light will always be limited to two very different directions (Fig.

Fig. 12.56

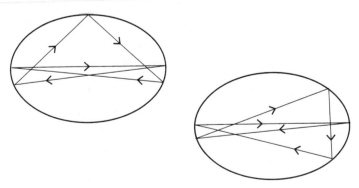

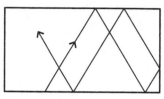

Fig. 12.57

12.57, left). Nor will it happen inside a circular billard table, because every straight segment of the path of the billiard ball (or the light ray), however many times it is reflected, will touch a smaller inside circle, and no ray will ever get inside that circle (Fig. 12.57, right). However, it can happen inside more complicated shapes, especially shapes part of whose boundary is straight. Figure 12.58 shows the simplest shape (called the *stadion*) with the property that a segment of a general reflected path will come arbitrarily close to any given straight path within the shape. It consists of two half-circles on the ends of a rectangle.

At this point, a possible *analogy* might strike us. Let's take the interval from 0 to 1 and wrap it round a circle, so that the 1 and 0 coincide, as in Fig. 12.59. We can now choose any rational number between 0 and 1 as our step length, and any starting point on the circle. Then we can imagine going round the circle in steps of the

Fig. 12.58

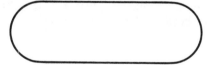

Fig. 12.59

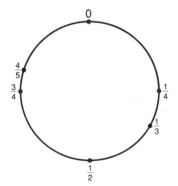

chosen length, passing 0 as often as necessary, and eventually getting back to where we started. For example, if we chose the number $\frac{5}{17}$, then after 17 steps, during which we would have gone round the circle exactly 5 times, we would return to our starting point. Thereafter we would only repeat our progress, like the 5-bounce path round the ellipse.

Suppose, however, that each of our steps was $\sqrt{2} - 1$ long. This is an irrational number, less than one. It is not possible that we shall ever return exactly to our starting point, because, if we did so, for example after taking N steps and circling the circle M times, then we would know that $\sqrt{2} - 1 = M/N$, and so $\sqrt{2}$ would be equal to the rational number $M/N + 1$. But we know that $\sqrt{2}$ is irrational.

So, we shall never repeat ourselves; but we can actually say more than that. We can say that sooner or later we shall get arbitrarily close to any position on the circle we choose. Suppose that we take the point $\frac{1}{4}$ on the circle, and decide to try to get to within 1/1,000,000 of it, on one side or the other. Will we do so, sooner or later? Yes, we will.

The behaviour of $\sqrt{2} - 1$, or any other irrational number (even if it is not less than 1), is reminiscent of the behaviour of a billiard that never repeats. Surprisingly, Nicole Oresme, the great fourteenth-century mathematician whom we have already met, appreciated the point about irrational numbers round the circle, which he expressed in his own vivid manner, thinking of planets circling the sun. In his *Tractatus de Commensurabilitate vel Incommensurabilitate Motuum Celi* he considers two bodies moving on a circle with steady but different velocities which are not a rational fraction of each other, and observes that 'No sector of a circle is so small that two such mobiles could not conjunct in it at some future time, and could not have conjuncted in it sometime [in the past].' Just so!

In tracing just some of the relatives and descendants of Heron's original problem, we have criss-crossed Cayley's mathematical landscape, and criss-crossed the history of mathematics. From Aristotle to Feynman, from mirrors to billiards, we find that every road has forks and turnings, that every solution suggests other problems, and that along the way the purest of intellectual puzzles are interwoven with the most practical and profound of applications. It is in the nature of mathematics as a human activity to be a scientific and artistic exploration of powerful and beautiful ideas. Fortunately, like chess, like music, this exploration can be enjoyed at every level, from the enthusiastic amateur to the world class grandmasters of mathematics whose names glow in the history books.

Solutions to problems, Chapter 12

12A: *The missing rectangle*

The centroid of the L shape lies on the line joining the centroid of the whole rectangle and the centroid of the 'missing' rectangle; or, to put that another way, the centroid of the whole rectangle divides the line joining the other two centroids in the ratio of their areas.

12B: *Rectangles or parallelograms?*

The centroid of a parallelogram is found by drawing both diagonals, just as for a rectangle, and the centroid of a 'tilting' L shape is found just as for a rectangular L shape, and so on.

12C: *Relative areas*

The centroid will divide the line joining the centres of the two rectangles in the inverse ratio of their 'weights', that is, their areas. So constructing their centroid with a straight edge also represents the ratios of their areas as a divided line segment. In Fig. 12.60, the ratio of the area of A to the area of B is QG/GP.

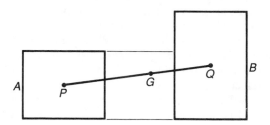

Fig. 12.60

12D: *Overlapping rectangles*

Yes. The centroid lies on the line joining their centres, and it also lies on the line joining the centre of the rectangle over which they overlap to the centroid of the shape formed by the outer boundary of the overlapping rectangles.

12E: *Rabbi Ben Moses' box*

Imagine two identical boxes which are the solution to Rabbi Ben Moses' problem. Placed open face to open face – which is the same as reflecting one in a mirror against its open face – they will form a closed solid which will have a maximum volume for its given surface area, and which will therefore be a cube. So each box separately will be half a cube, which was his conclusion.

12F: *The maximum isosceles triangle*

Reflect the triangle in its base, and you get a rhombus whose area will be greatest when it is inscribed in a circle, that is, when it is a square (Fig. 12.61). Therefore, the isosceles triangle has a maximum area when it is half a square and the top vertex angle is a right angle. If you treat one arm of the angle as the base, and allow the other arm to move, then it is also clear that its area is a maximum when the arms are perpendicular, because the base is fixed and the variable height is then a maximum (Fig. 12.62).

Fig. 12.61

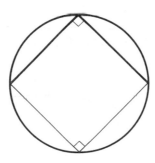

Fig. 12.62

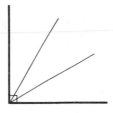

12G: *The advantageous race*

Not so surprisingly, the solution once again depends on reflection: reflect Jack's position in the line of the path, and draw a line through the reflected point and Jill's position to cut the path (Fig. 12.63). That is where Jack's advantage over Jill will be greatest. Jill on the other hand would prefer to run to the point where the perpendicular

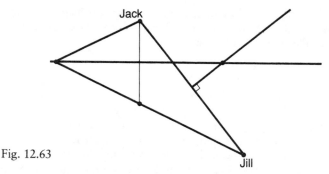

Fig. 12.63

bisector of the line joining her to Jack cuts the path, because then they will run the same distance.

12H: *Jack's puzzle, revised*

Consider an ellipse whose foci are A and B. If P is any point on the ellipse, then $AP + PB$ is constant. Therefore, if we choose the smallest ellipse with foci A and B that just touches the curve of the river, and it touches the river at P, then $AP + PB$ will be a minimum.

Fig. 12.64

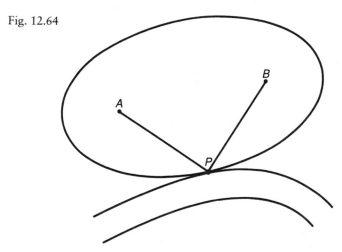

12I: *Weighted routes*

The traffic between B and C is double density; so, instead of trying to minimize $AX + BX + CX$, we should be trying to minimize

$$(AX + XB) + 2(BX + XC) + (CX + XA) = 2AX + 3BX + 3CX.$$

If we now apply exactly the same argument as before, imagining that when A is replaced by A' there is a new minimum position at X', then we reach the same absurd conclusion as before; if we try replacing B with a new point on BX labelled B', and then apply the same argument, the same conclusion also applies.

Therefore, the angles will be unchanging in this case also: the difference is that they will not be 120° and we cannot spot them at a glance. However, we can consider an easy case as in this figure, where the towns form an equilateral triangle, and the minimum point will lie on the line of symmetry (Fig. 12.65). The total distance is

$$2AX + 3BX + 3CX = 2(\sqrt{3} - x) + 3\sqrt{(1 + x^2)} + 3\sqrt{(1 + x^2)},$$

which is a minimum (by calculus) when $x = \frac{1}{4}\sqrt{2}$ and the angles between AX, BX and CX are approximately 141°, 109½° and 109½°.

Fig. 12.65

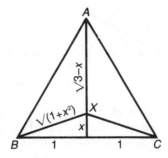

12J: *The Steiner network for a square*

We construct equilateral triangles on a pair of opposite sides, and draw their circumcircles. It is convenient to call the side of the square 2 units, and the height of each equilateral triangle is then $\sqrt{3}$ (Fig. 12.66). Therefore the minimum Steiner network, which is the length of PS, is $\sqrt{3} + 2 + \sqrt{3} = 2 + 2\sqrt{3}$.

Fig. 12.66

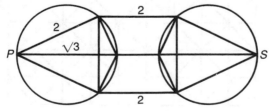

12K: *An altitude property*

Let O be the orthocentre where the altitudes meet. Then $\widehat{OPC} = \widehat{OQC} = 90°$, and so P and Q lie on the circle on OC as diameter (Fig. 12.67). Therefore we can use the 'angle in the same segment' property to conclude that the angles $\widehat{QPC}$ and $\widehat{QOC}$ are equal. However, $AROQ$ is also a cyclic quadrilateral and so we can conclude that $\widehat{QOC} = 180° - \widehat{QOR} = A$. Similarly, $\widehat{BPR} = \widehat{BOR} = A$.

Fig. 12.67

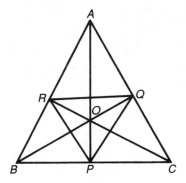

12L: *A billiards puzzle*

By simply drawing a rectangle 9 by 7, we can follow the path of the ball and see that it ends up in the top right-hand pocket (Fig. 12.68). We could also argue as follows, without drawing: for every 9 diagonal

Fig. 12.68

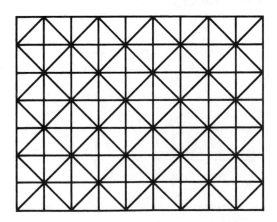

units that the ball moves, it will go from one short side of the table to the other. For every 7 diagonal units that it moves, it will move from one long side to the other. To end up in a pocket, it must have covered the length of the table and its width a whole number of times each. Therefore we find the lowest common multiple of 9 and 7, which is $9 \times 7 = 63$, and we can predict that after 63 diagonal moves it will be in a pocket. Since it will have covered the length of the table 7 times and the width of the table 9 times (both odd numbers), it will be at the opposite end and on the opposite side from the corner where it started. Therefore it ends up in the *opposite* corner.

Chapter 13

A mathematical adventure

This chapter has the format of an adventure game. On each page there are one or more numbered frames, starting with Frame 1. In each frame there is some text, followed either by an instruction to turn to another named frame, or a description of two or more possible trains of thought at that point, each leading to a named frame. Your choice of which train of thought to follow determines which frame you turn to next. *The frame numbers are printed at the top of each box to help you jump straight to the next frame without being distracted by the material in between.* So that you can keep track of the route you have taken, a plan of all the frames, showing how they are linked together, is in Fig. 13.1.

If you did not wish to actually do any mathematics yourself, you can skim through the maze anyway, following the trail, backtracking, taking whatever turnings appeal to you, and so exploring this portion of the mathematical landscape. But we hope that at each stage you will enjoy attempting to follow, and work on, the train of thought that you have chosen, trying to anticipate the contents of the next frame. However you treat this maze, the resulting journey will be a good model of the processes of exploring a mathematical problem.

The questions and ideas presented here are only a small fraction of those that might be asked, and which could occur to you. Some frames may simply tell you that if you wish to pursue a particular path any further – then *you are on your own*! Of course, that is also typical of mathematical exploration, in which it is so easy (and enjoyable) to leave the familiar landscape behind and set out in your own chosen direction. There is no end point to the exploration, though there are several frames at which you might decide to call it a day – but such a decision will be yours alone, based on your feeling for what you have discovered, and your judgement of the remaining options. To start the adventure, turn to Frame 1 on the next page.

To avoid being distracted by the right-hand page while reading a left-hand page, or vice versa, it is advisable to place a blank piece of paper between them, as you read.

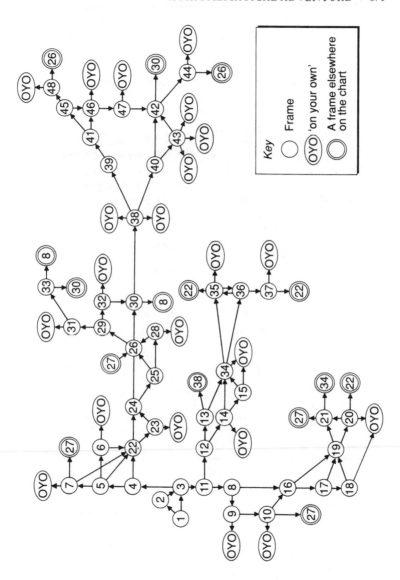

Fig. 13.1

1

This puzzle appeared in Edward Riddle's edition (1840) of the *Recreations in Mathematics and Natural Philosophy* of Jacques Ozanam, where it was attributed to Count Buffon, the eminent French naturalist and translator of Newton's *Principia*.

Given any irregular polygon, the midpoints of the sides are joined in sequence, and this process is repeated, again and again. 'It is required to find the point where these divisions will terminate.' It is very natural to wonder what happens if you do something again and again ... and again. Such problems also have a ready appeal to modern mathematicians who can use computers to do experiments and hence investigate the results scientifically before they try to prove anything. Fortunately, a computer is not necessary to study Count Buffon's problem. Figure 13.2 shows an original polygon, and then the first and second stages in the process.

Fig. 13.2

- You spot the reason why Buffon was certain that the process did tend to a point – **turn to Frame 3**.

- You want a hint towards Buffon's explanation – **turn to Frame 2**.

2

Buffon considered how the *centroid* (centre of gravity) of the original vertices changes (or does not change) when the vertices are replaced by the midpoints of the sides.

● **Turn to Frame 3.**

3

Buffon noticed that the centroid of all the midpoints was the same as the centroid of the original vertices. Therefore this centroid never changes, and Buffon assumed without further argument that every sequence of polygons did indeed tend to the original centroid.

Taking a pentagon as an example, the centroid of A, B, C, D and E is $\frac{1}{5}(A + B + C + D + E)$; (here we are imagining that each point can be described by coordinates, which we can add and average). The midpoints of the sides are $\frac{1}{2}(A + B)$ and so on, and the centroid of the midpoints is

$$\frac{1}{5}[\frac{1}{2}(A + B) + \frac{1}{2}(B + C) + \frac{1}{2}(C + D) + \frac{1}{2}(D + E) + \frac{1}{2}(E + A)]$$
$$= \frac{1}{5}(A + B + C + D + E),$$

the same point.

● You wonder how fast the process will tend to a point – **turn to Frame 4**.

● You wonder what other similar processes might be applied to a polygon – **turn to Frame 11**.

4

Buffon's process applied to any triangle (Fig 13.3), results at each step in a triangle which is one-quarter the area of the original triangle. By taking enough steps we can make the Nth triangle as small as we choose, and so the limit of the sequence will be a point. A square, or a general quadrilateral, follows a similar pattern (Fig 13.4), except that the area is reduced by one half at each stage.

Fig. 13.3

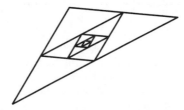

Fig. 13.4

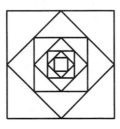 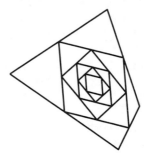

- You wonder how fast the size of a hexagon will be reduced in area – **turn to Frame 5**.

- You wonder how the shape of a hexagon changes as a result of this process, having observed that they seem to become more 'regular' – **turn to Frame 22**.

5

To estimate the reduction in the polygon after applying Buffon's process once, it helps to recall that, if two midpoints of the sides of a triangle are joined, the resulting small triangle is one quarter the area of the original triangle. In the hexagon shown in Fig. 13.5, we have cut off three small triangles by three midpoint lines. The shaded areas are exactly one quarter the total area of the three triangles, excluding the middle triangle. Figure 13.6 shows the same hexagon, with the other three midpoint lines drawn: once again, the shaded triangles are one quarter of the three triangles excluding the centre. If we were confident that in at least one of the diagrams the three outer triangles were at least as large as the inner triangle, then one quarter of their total area would be at least $\frac{1}{8}$ of the area of the hexagon, and so the area would be reduced by at least $\frac{1}{8}$ at each stage.

Fig. 13.5

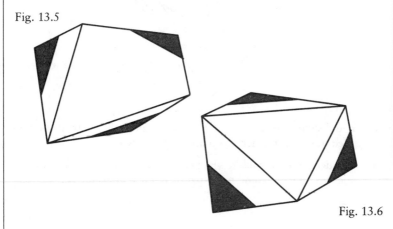

Fig. 13.6

- The thought occurs to you that, if the hexagons do become more 'regular', then it would be much easier to estimate the rate at which the area decreases – **turn to Frame 22.**

- You wonder what happens to a pentagon – **turn to Frame 6.**

- You suppose that, if the sides were divided in the ratio (say) of $1:2$, then the area would decrease even more slowly – **turn to Frame 7.**

6

Figure 13.7 shows the fate of one pentagon after 5 stages. Naturally, the 'opposite sides' are not more nearly parallel, since a pentagon does not have opposite sides: yet it seems that it is 'more regular'. You realize that each side and its opposite *diagonal* have become more nearly parallel – and that this is a kind of analogue of the opposite sides of an even-number-sided polygon becoming more parallel.

Fig. 13.7

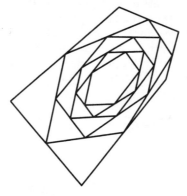

- You decide to investigate the pentagon and other higher-number-sided polygons further – **you are on your own.**

- You decide to go back to the hexagon and study why its opposite sides become more and more parallel – **turn to Frame 22.**

7

An experiment shows very clearly what you suspected: with sides divided in ratios such as 1:2, the limit is approached more slowly (Fig. 13.8). This suggests that Buffon's process works fastest when the sides are divided at their midpoints.

Fig. 13.8

- You wonder whether the rate at which the area decreases actually increases with each step, or whether perhaps it decreases – **you are on your own.**

- You wonder what happens when the same processes are done backwards – **turn to Frame 27.**

- You decide to return to the investigation of the hexagon, on the grounds that it is an elegant shape and there are probably some elegant arguments to fit the case – **turn to Frame 22.**

8

Buffon's argument about centroids, and the proof by averaging adjacent vertices, does not assume anywhere that the points lie in a plane. Indeed, one of the advantages of this use of averages is that it applies to any number of dimensions, and in particular to three dimensions, so a skew polygon will also tend to a point. The thought occurs to you that a skew quadrilateral will at the very first stage produce a plane parallelogram, as in Fig. 13.9.

Fig. 13.9

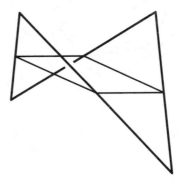

- This suggests the question: do all skew polygons tend to become less skew as a result of this process? – **turn to Frame 9.**

- You wonder what would happen if you applied a similar kind of process, not to a polygon but to a polyhedron, such as a cube or tetrahedron – **turn to Frame 16.**

9

Figure 13.10 can be seen in two ways, either as a rather irregular heptagon, or as a picture of a skew heptagon. From the first point of view, it suggests that an irregular plane polygon becomes rather less irregular as a result of Buffon's process, in the sense that it is closer to being convex and its angles are more nearly equal. From the second perspective, Buffon's process does seem to make the skew octagon less skew, and any simple three-dimensional model, composed for example of straws or matchsticks, will confirm this observation.

Fig. 13.10

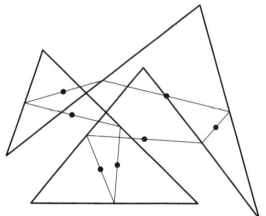

- You decide to do an experimental test of the idea that irregular polygons become more regular – **turn to Frame 10.**

- You want to investigate further how skew polygons tend to a plane form – **you are on your own.**

10

Figure 13.11 shows one experimental test of Buffon's process on Fig. 13.10. The midpoints in the first figure are the vertices of the outer polygon in the second figure.

Fig. 13.11

- The sequence of figures seems to be tending to an elliptical shape, rather than becoming regular. If you wish to pursue this line of enquiry further – **you are on your own.**

- Reading the sequence of diagrams backwards suggests to you that Buffon's process when reversed will turn a 'normal' looking hexagon into something extremely irregular – **turn to Frame 27.**

- To see what happens when Buffon's idea is applied to polyhedra – **turn to Frame 16.**

11

Buffon's original problem suggests several natural variations. His observation that the midpoints of the sides are the centroids of adjacent vertices suggests that you might consider other centroids. Forgetting centroids, there are many other ways to construct new points based on pairs of vertices: and there is also the question whether the original polygon needs to be a plane polygon at all.

- You decide to investigate the results of finding the centroids of sets of three adjacent vertices: when you have done so – **turn to Frame 12.**

- You investigate the same process starting with a skew polygon (one whose vertices do not lie in a plane) – **turn to Frame 8.**

12

By experiment, when the centroids of sets of three adjacent vertices are marked, any irregular hexagon at once produces a hexagon whose opposite sides are equal and parallel (Fig. 13.12).

Fig. 13.12

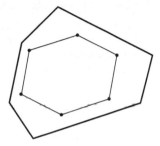

Fig. 13.13

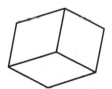

- You decide to prove this result by means of the usual methods – **turn to Frame 13.**

You spot that the shape in Fig. 13.12 is much like the outline of a box. In fact, it can be turned into a box by completing three parallelograms, as in Fig. 13.13.

- You wonder what happens when you take an irregular hexagon and complete six parallelograms inside it – **turn to Frame 14.**

13

The centroids can be calculated by thinking of the vertices as the ends of vectors, and averaging the points, almost as if they were numbers (Fig. 13.14). A' is $\frac{1}{3}(F + A + B)$ and B' is $\frac{1}{3}(A + B + C)$. The vector $A'B'$ is therefore

$$\tfrac{1}{3}(A + B + C) - \tfrac{1}{3}(F + A + B) = \tfrac{1}{3}(C - F).$$

But F and C are diagonally opposite vertices of the original hexagon, and so, by symmetry, the vector $D'E'$ will be $\frac{1}{3}(F - C)$. In other words, $A'B'$ and $D'E'$ are equal displacements, except for the change of sign. They are therefore equal in length and parallel, which is what we wanted to prove.

Fig. 13.14

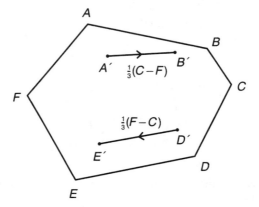

- You decide to investigate other ways of constructing 'derived' polygons – **turn to Frame 34.**

- It occurs to you that, instead of finding averages of geometrical points, as if they were numbers, you could actually be finding the averages of numbers, for example by writing down a circle of numbers, averaging them in pairs, and repeating the process – **turn to Frame 38.**

14

Experiment shows that, if any three consecutive vertices of an irregular hexagon are taken as the vertices of a parallelogram, and the fourth vertex is added to complete the parallelogram, then the six new vertices form a picture of a prism, as in Fig. 13.15.

Fig. 13.15

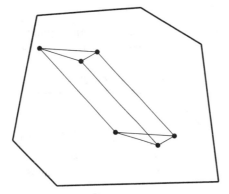

- You decide to prove this – **turn to Frame 15.**

- You decide to investigate similar constructions for octagons and other polygons with more sides – **you are on your own.**

- You decide to look at other ways of constructing new polygons from old – **turn to Frame 34.**

15

If A, B and C are three consecutive vertices, then the vector from B to C is $C - B$, and so the vector from A to the missing vertex of the parallelogram, D, is also $C - B$ (Fig. 13.16). Therefore, D is the point $A + C - B$. In Fig. 13.17, the six new vertices are as marked. The three lower vertices are all displaced from the three upper vertices by the vector

$$B + F + D - A - C - E,$$

and so the figure does indeed consist of three points, and their copies by the same parallel displacements; that is, it is a picture of a 'prism'.

Fig. 13.16

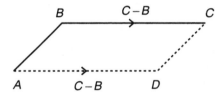

Fig. 13.17

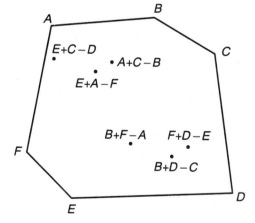

- You decide to investigate similar constructions for octagons and other polygons with more sides – **you are on your own**.

- You decide to look at other ways of constructing new polygons from old – turn to **Frame 34**.

16

A skew quadrilateral looks much like a tetrahedron, and becomes a tetrahedron when its diagonals are added (Fig. 13.18).

Fig. 13.18

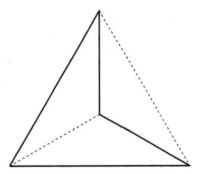

- This suggests that Buffon's process for a tetrahedron could mean marking and joining all six of the midpoints of its edges – **turn to Frame 17.**

- On the other hand, finding the centroid of adjacent vertices at either end of an edge of a polygon seems analogous to finding the centroid of the vertices round the edge of a polyhedron, so you decide that Buffon's process for a polyhedron should consist of marking and joining the centres of the faces – **turn to Frame 19.**

17

Marking and joining the midpoints of adjacent edges produces the octahedron shown in Fig. 13.19.

Fig. 13.19

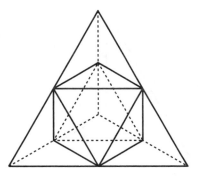

- You decide to repeat the same process for this octahedron – **turn to Frame 18.**

- This seems to be becoming too complicated, so you go back to the idea of marking the centres of the faces of a polyhedron – **turn to Frame 19.**

18

Joining the midpoints of adjacent edges produces a *cuboctahedron* (Fig. 13.20) with six square faces and eight triangular faces. This process is equivalent to truncating the polyhedron: in other words, slicing off each vertex, in this case as far as the midpoints of the edges.

Fig. 13.20

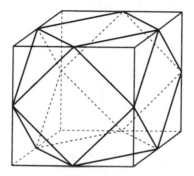

- If you want to investigate the results of truncating this cuboctahedron – **you are on your own.**

- You decide to join the centres of the faces of the original tetrahedron – **turn to Frame 19.**

19

A tetrahedron has the same number (four) of faces and vertices; so it is no surprise that, if the centre of each face becomes the vertex of a new solid, as in Fig. 13.21, it is also a tetrahedron. The cube, on the other hand, has as many faces and vertices as the octahedron has vertices and faces respectively: in fact, the centres of the faces of a cube are the vertices of a regular octahedron (Fig. 13.22) – and if the same process is applied to the regular octahedron, then you get back to a cube.

Fig. 13.21

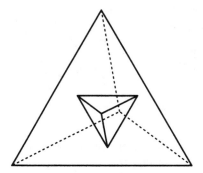

Fig. 13.22

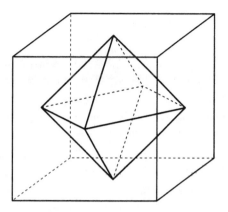

- You wonder why a pair of polyhedra in three dimensions should switch one to the other and back, whereas a plane polygon just becomes another plane polygon with the same number of sides – **turn to Frame 20.**

- You wonder what other kinds of polyhedra have as many vertices as they have faces, so that they might share the property of the tetrahedron of being transformed into themselves – **turn to Frame 21.**

20

Any geometer or topologist will confirm that there is no simple relationship between the properties of figures in spaces having a different number of dimensions. In particular, each of the euclidean spaces of two, three and four dimensions has many features which are not found in the other two, and which are not found in spaces of higher dimensions. So the noted difference is typical, rather than surprising.

- You decide to look for other striking differences between polygons and polyhedra – **you are on your own.**

- You go back to the problem of polyhedra with equal numbers of vertices and faces – **turn to Frame 21.**

- You go back to the question of polygons becoming more regular through Buffon's process, and investigate what happens to a plane hexagon – **turn to Frame 22.**

21

By Euler's relationship for any 'ordinary' polyhedron, $V + F = E + 2$. So, if $V = F$, then $2V = 2F = E + 2$, and therefore E must be even. The cases $E = 2$ and $V = 2$, and $E = 4$ and $V = F = 3$, cannot be constructed as polyhedra. For $E = 6$ and $V = F = 4$, we get the tetrahedron that we have already considered. The case $E = 8$ and $V = F = 5$ corresponds to a square pyramid (Fig. 13.23). This joins on to the previous case because a tetrahedron can be thought of as a triangular pyramid. The next case, $E = 10$ and $V = F = 6$ could yield a pentagonal pyramid – and so on – all higher values of E (and the matching equal values of V and F) can be matched against pyramids with more and more sides.

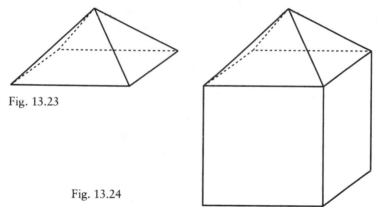

Fig. 13.23

Fig. 13.24

In each case, joining the centres of the faces produces another pyramid with the same number of faces. However, there are other possibilities, such as a pyramid on a cube (Fig. 13.24), with $E = 16$ and $V = F = 9$. The joins of the centres of the faces and centre of the base form a polyhedron which is a distorted version of the original solid.

- The process of joining the centres of the faces can be reversed: you can go backwards, as it were. You decide to investigate Buffon's process backwards – **turn to Frame 27.**

- You abandon polyhedra and go back to other ways of constructing new polygons from old – **turn to Frame 34.**

22

The hexagon shown in Fig. 13.25 is not strikingly irregular to start with – yet, after a few stages of Buffon's process, it is markedly more regular. In particular, opposite edges are more nearly parallel.

Fig. 13.25

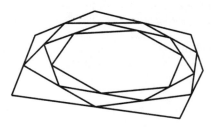

- You wonder what 'more regular' might mean for a polygon: it seems about time that this vague expression was made clearer and less ambiguous – **turn to Frame 23**.

- You decide to try to prove that the hexagon is indeed in some sense becoming 'more regular' – **turn to Frame 24**.

23

Deciding what an expression such as 'more regular polygon' might mean is a matter of experience, of discovery, and cannot be decided in advance: in the present case, experiment suggests that opposite sides of the hexagon become more parallel, and more nearly equal in length, though adjacent sides may still differ greatly in length and the angles between them may not be that close to 120°. So we are thinking in terms of a hexagon which looks much like the one in Fig. 13.26.

Fig. 13.26

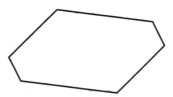

- If you want to pursue this question of definition – **you are on your own.**

- You decide to prove that the hexagon becomes, in some sense, 'more regular' – **turn to Frame 24.**

24

Figure 13.27 shows an original hexagon, and the result of applying Buffon's process twice. The third hexagon is clearly less irregular than the original hexagon. To see why this is so, we calculate the vertices of the third hexagon, and then calculate the vectors representing its sides and compare them with the sides of the original hexagon, as in Fig. 13.28. Calling the original vertices A to F, the vertices of the second hexagon are $\frac{1}{2}(A + B)$, $\frac{1}{2}(B + C)$ and so on until $\frac{1}{2}(F + A)$. When we average consecutive second vertices we find that the vertices of the third hexagon are $\frac{1}{4}(A + 2B + C)$, $\frac{1}{4}(B + 2C + D)$ and so on, round to $\frac{1}{4}(F + 2A + B)$. Now we calculate the vectors for the

Fig. 13.27

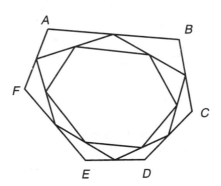

Fig. 13.28

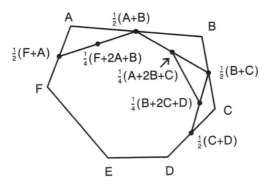

Fig. 13.29

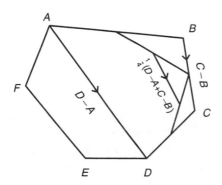

Fig. 13.30

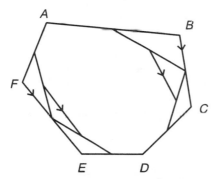

sides (Fig. 13.29). The original vector $\overrightarrow{BC}$ is represented by $C - B$. The vector from $\frac{1}{4}(A + 2B + C)$ to $\frac{1}{4}(B + 2C + D)$ is

$$\frac{1}{4}(B + 2C + D - A - 2B - C) = \frac{1}{4}(D - A + C - B).$$

Looking at this expression, we see that it is one quarter of the sum of $C - B$ and $D - A$, where $C - B$ is the original side vector from B to C, and $D - A$ is the original diagonal vector from A to D. This tells us something very important, because we know that, when vectors are added, the direction of their sum lies *between* the directions of the original vectors. In other words, the side of the third hexagon which corresponds to the original side BC makes a smaller angle with the original diagonal AD.

Since it will also be true that the side of the third hexagon corresponding to the original side FE (Fig. 13.30), which was opposite BC, also makes a smaller angle with AD, we see that BC and FE have

(continued)

been transformed, in two steps, into lines that are more nearly parallel to each other. (There is a gap in this argument, which we will not fill here: the argument is correct in principle.)

Here is another way to look at what is happening: the original difference between the sides BC and FE, regarded as vectors, was $(E - F) - (C - B)$. The difference between their corresponding sides in the third hexagon is $\frac{1}{4}(D - A + E - F) - \frac{1}{4}(D - A + C - B) = \frac{1}{4}(E - F) - \frac{1}{4}(C - B)$. In other words, the difference between them, regarded as a vector, is one quarter of what it was before. After two more stages, it will be $\frac{1}{16}$ of what it was originally, and so on. By constructing enough stages, we can make the difference as small as we choose, which is another way of saying that the sides eventually are represented by vectors whose difference is arbitrarily small – the sides are very nearly parallel and equal in length.

- You spot a relationship between the two sets of alternate midpoints of the original hexagon – **turn to Frame 25**.

- It occurs to you that if Buffon's process makes a hexagon more and more 'regular', or pseudo-regular, then the reverse of Buffon's process will make it more irregular – **turn to Frame 26**.

25

One set of alternate midpoints is $\frac{1}{2}(A + B)$, $\frac{1}{2}(C + D)$ and $\frac{1}{2}(E + F)$. The other set is $\frac{1}{2}(B + C)$, $\frac{1}{2}(D + E)$ and $\frac{1}{2}(F + A)$. The average of either set, which is to say the centroid of either triangle, is the same: $\frac{1}{6}(A + B + C + D + E + F)$ which is also the centroid of the original hexagon.

- It occurs to you that *all* the hexagons resulting from Buffon's process will have this property, except for the original hexagon which can be chosen at random. You wonder whether Buffon's process can be reversed *only* for hexagons with this alternate-vertices property, or whether it can be reversed for any arbitrarily chosen hexagon – **turn to Frame 28**.

- You decide to see whether Buffon's process when reversed makes the hexagons less regular – **turn to Frame 26**.

26

Given five of the six vertices of a hexagon, you can reverse Buffon's process, starting with any point you choose, because the starting point and the sixth point will then uniquely define the final, sixth, vertex as the midpoint of the line segment needed to close the hexagon. In Fig. 13.31, the five vertices are labelled A to E. The first arbitrarily chosen point is X, and the sixth point is Y. Here F is the midpoint of XY and the two sets of three points, A, C, E and B, D, F have the same centroids. This means that any point X′ will now lead to a closed hexagon, as Fig. 13.32 illustrates.

Fig. 13.31

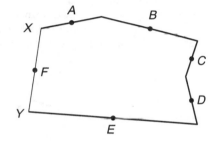

Fig. 13.32

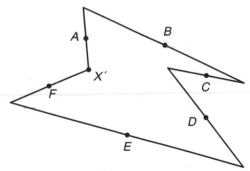

- You wonder what happens if you try to work Buffon backwards, starting with an arbitrary triangle – **turn to Frame 29**.

- You decide to see what happens to a parallelogram, knowing that the midpoints of every quadrilateral form a parallelogram, so to work Buffon backwards for a quadrilateral, it must be a parallelogram in the first place – **turn to Frame 30**.

27

Reversing Buffon's process is a tempting possibility, but it raises several questions: when will it work? When will the new polygon be closed? How regular or irregular will it be? What variety of polygons can be created by working backwards? Figure 13.33 shows a set of six points which are taken as the midpoints of a hexagon constructed by working backwards, starting with the new vertex X. However, the six midpoints were constructed in the first place by marking the midpoints of a hexagon, so it is not so surprising that we can work backwards from them successfully. Figure 13.34 illustrates an attempt to construct a pentagon by working backwards, taking the five given points as dividing the new sides in the ratio of $2:1$ (rather than being the midpoints). The result is a failure – the point marked M ought to divide the side XV in the ratio $2:1$, but it does not even lie on the line XV.

Fig. 13.33

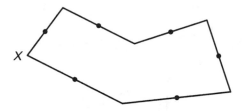

Fig. 13.34

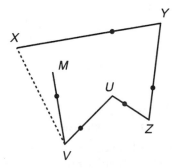

● To investigate these problems for a hexagon – **turn to Frame 26**.

28

Figure 13.35 shows the result of taking six points, supposing that they are the midpoints of another hexagon, and taking an arbitrary point, X, as one of the new vertices. After constructing a sequence of new sides and vertices, with the given points as the midpoints of the sides, the hexagon does not close, so there is something wrong.

Fig. 13.35

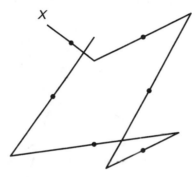

- You decide to look at Buffon's process reversed for general polygons – **turn to Frame 26.**

- You decide to discover for which points X the hexagon *does* close – **you are on your own.**

29

By experiment, given three points as the midpoints of the sides of a triangle, if we start with an arbitrary point and work backwards, the triangle does not close. In Fig. 13.36, we start at X and can construct Y, Z and then X' in sequence. The point X' is some way from X. However, if we 'go round twice', as it were, then it does close: the next three points are Y', Z' and then our original point X.

Fig. 13.36

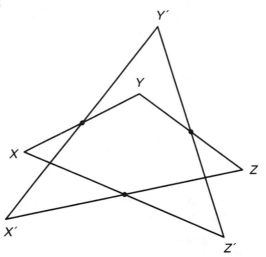

- You decide to investigate this result by calculation – **turn to Frame 31**.

- You recall a property of the triangle which indicates immediately the starting points for which the triangle will close – **turn to Frame 32**.

30

Figure 13.37 shows a parallelogram, and a quadrilateral whose mid-points are the parallelogram's vertices. As the second diagram illustrates, there are any number of quadrilaterals with the same four midpoints.

Fig. 13.37

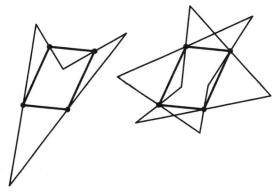

• It occurs to you that, in finding averages of geometrical points, you are effectively finding the averages of their x and y coordinates separately, so you could just as well consider Buffon's process applied to one-dimensional chains of numbers – **turn to Frame 38.**

• You decide to explore another variant – points in three-dimensional space – for which you will effectively be finding the averages of three separate coordinates at a time – **turn to Frame 8.**

31

Let the three given points be A, B and C, and the arbitrarily chosen vertex, X (Fig 13.38). If A is the midpoint of XY, then $Y - A = A - X$ and so $Y = 2A - X$. Similarly the third new vertex Z will be $2B - (2A - X) = 2B - 2A + X$. The fourth new vertex, which will be identical to X when the triangle closes, will be

$$2C - (2B - 2A + X) = 2C - 2B + 2A - X.$$

The result of going round a second time will be that you arrive at the point

$$2C - 2B + 2A - (2C - 2B + 2A - X) = X,$$

which explains why going round twice is successful.

If we go round just once, $2C - 2B + 2A - X$ will indeed be equal to X only if $X = 2C - 2B + 2A - X$, or $X = C - B + A$. What does this mean? It can be read in one way as $C + (A - B)$, which is the point X on the left-hand diagram of Fig. 13.39. On the other hand, it can also be read as $A + (C - B)$, which happens to be the same point – the fourth vertex of a parallelogram with vertices A, B and C.

Fig. 13.38

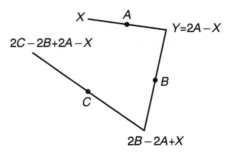

Fig. 13.39

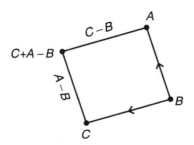

If we had started with a point X and supposed that B or C were the first midpoints (not A), then we have discovered two other points from which the triangle closes, as in Fig. 13.40. We discover that the triangle only closes when we start with one of the vertices of the triangle similar to ABC but with double the length of side. This is just another way of recalling the fact that the triangle formed by the midpoints of the sides of a triangle is similar to the triangle, but half the linear size.

Fig. 13.40

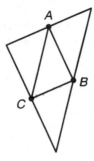

- It occurs to you that, if you had started with *five* points instead of three, and a chosen starting point, X, then the condition that the last point coincided with X would be (following the pattern above) $2E - 2D + 2C - 2B + 2A - X = X$, so a pentagon will only close if $X = A - B + C - D + E$ (or the similar expression starting with a different letter) – **you are on your own.**

- It occurs to you that, if you started with *four* points, then the condition would be $2D - 2C + 2B - 2A + X = X$, which simplifies to $D - C + B - A = 0$ which does not depend on X. You decide to investigate what this means – **turn to Frame 33.**

32

You realize that, if the original points are A, B and C, and the triangle does close, then you will get a figure like Fig. 13.40 in which A, B and C are the midpoints of the sides, and the triangle ABC is the same shape as the larger triangle but half the linear size. Therefore, the only three points for which the triangle will close will be the vertices of this larger triangle.

- You decide to look for matching properties of the midpoints of pentagons and other polygons – **you are on your own.**

- You recall that the midpoints of the sides of a quadrilateral have a simple property – they form a parallelogram – **turn to Frame 30.**

33

$D - C + B - A = 0$ can be written $D - C = A - B$ (Fig. 13.41). This means that the four points must form a parallelogram, because a pair of sides is equal in length and direction. (The same equation can also be written $D - A = C - B$, so it also says that the other pair of sides is parallel and equal in length.)

Fig. 13.41

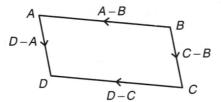

- You go back to the question of reversing Buffon for a parallelogram – **turn to Frame 30.**

- It occurs to you that these calculations by adding and subtracting points will work just as well in three dimensions – turn to **Frame 8.**

34

Starting with a given polygon, there is no reason why the new vertices should actually be on the edges of the polygon – and we have already seen an example of a construction in which they were not. Figure 13.42 is the diagram for Napoleon's theorem, in which the three new points are the centres of equilateral triangles constructed outwards on the edges. Alternatively, they can be seen as the vertices of triangles constructed on the sides with angles of 30°, 30° and 120°.

Fig. 13.42

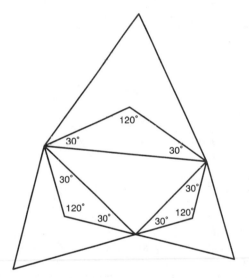

- You decide to try a different shape of triangle: isosceles right-angled triangles fitted outwards on the sides of an arbitrary triangle – **turn to Frame 35.**

- You decide to move from equilateral triangles on a triangle to squares fitted outwards on the sides of a quadrilateral – **turn to Frame 36.**

- You try triangles fitted outwards on the sides of a pentagon or hexagon – **you are on your own.**

35

Figure 13.43 shows three half-squares constructed on the sides of an arbitrary triangle, *ABC*. By experiment, we can discover surprising facts about this figure which are reminiscent of (though not the same as) the properties of the figures for Napoleon's theorem and the Fermat point theorem:

Fig. 13.43

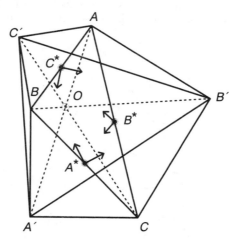

The lines *AA'*, *BB'* and *CC'* concur, though not at angles of 60°.

The lines *AA'* and *B'C'* are equal in length and perpendicular to each other; so are the pairs *BB'* and *C'A'*, and *CC'* and *A'B'*.

If *A** is the midpoint of *BC*, then *C'A*B* is also an isosceles right-angled triangle – a half-square. So are triangles *A'B*C'* and *B'C*A'*.

- Having succeeded with this idea, you decide to look at squares fitted outwards on the sides of a quadrilateral – **turn to Frame 36.**

- The appearance of half-squares following a construction using half-squares (and the appearance of equilateral triangles in the Napoleon figure) reminds you of the tendency for more-or-less regular hexagons to appear as a result of Buffon's process – **turn to Frame 22.**

- You decide to investigate this and related figures further – **you are on your own.**

36

This figure shows squares on the sides of an arbitrary quadrilateral. Just as for the Napoleon figure, the centres of the added shapes have been marked, and experimental measurement will show that *PR* and *QS* are equal in length and perpendicular, but there do not seem to be any other properties that are easy to spot.

Fig. 13.44

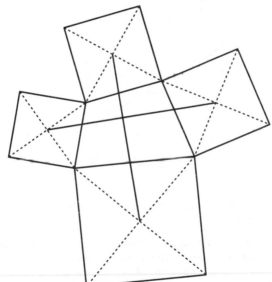

- Perhaps the arbitrary quadrilateral is *too* arbitrary: you decide to see what happens if the original quadrilateral is rather more 'regular', say, if it is a parallelogram – **turn to Frame 37**.

- You decide to try a different shape of triangle, on a triangle, after all: isosceles right-angled triangles fitted outwards on the sides of an arbitrary triangle – **turn to Frame 35**.

37

If the original quadrilateral is a parallelogram, then the centres of the added squares form another square (Fig. 13.45). Moreover, the triangles APS, BQP, CRQ and DSR are all congruent to each other.

Fig. 13.45

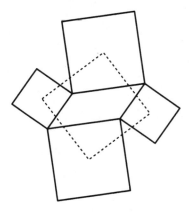

- You decide to investigate this and related figures further – **you are on your own.**

- The appearance of a square following a construction using half-squares reminds you of the tendency for more-or-less regular hexagons to appear as a result of Buffon's process – **turn to Frame 22.**

38

Fig. 13.46

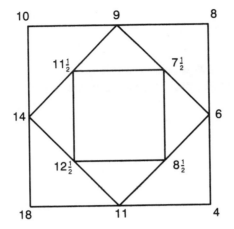

You write four numbers in a square, as in Fig. 13.46, and insert the first square of averages of consecutive pairs. But there is hardly any room for the next stage, so you stop, and rewrite them like this, simply remembering that the left-hand and right-hand ends are meant to be joined together.

10		8		4		18	
	9		6		11		14
$11\frac{1}{2}$		$7\frac{1}{2}$		$8\frac{1}{2}$		$12\frac{1}{2}$	
	$9\frac{1}{2}$		8		$10\frac{1}{2}$		12
$10\frac{3}{4}$		$8\frac{3}{4}$		$9\frac{1}{4}$		$11\frac{1}{4}$	
...		...		...		...	

The repeated averages seem to settle down rather quickly to about the same quantity.

● You want to consider the connection between this quantity and the original numbers – **turn to Frame 39.**

(continued)

- You wonder *how fast* they are settling down: Can this be measured? What sets of numbers will settle down fastest? Slowest? – **turn to Frame 40.**

- You wonder what happens if you start with a different number of numbers – **you are on your own.**

- You decide that, to be a good analogue of the geometrical case of a convex polygon, you should have a chain of numbers which rises and then falls, like this: 0 4 16 20 12 10 – **you are on your own.**

39

If the original numbers are a, b, c and d, then the next row is $\frac{1}{2}(a + b)$, $\frac{1}{2}(b + c)$, $\frac{1}{2}(c + d)$ and $\frac{1}{2}(d + a)$, and the sum of these four numbers is the same as the sum of the original numbers. The average of each row is therefore constant, and as the numbers get closer together, they will all tend to this average. In our example, that average is 10.

- You wonder how quickly this average is approached – **turn to Frame 40.**

- You wonder if the repeats would settle down faster if you averaged three consecutive numbers at a time – **turn to Frame 41.**

40

By using algebra, instead of actual numbers, we can observe the pattern of the calculation, just as we could in the geometrical case:

a	b	c	d
$\frac{1}{2}(a + b)$	$\frac{1}{2}(b + c)$	$\frac{1}{2}(c + d)$	$\frac{1}{2}(d + a)$
$\frac{1}{4}(d + 2a + b)$	$\frac{1}{4}(a + 2b + c)$	$\frac{1}{4}(b + 2c + d)$	$\frac{1}{4}(c + 2d + a)$
$\frac{1}{8}(d + 3a + 3b + c)$	$\frac{1}{8}(a + 3b + 3c + d)$	$\frac{1}{8}(b + 3c + 3d + a)$	$\frac{1}{8}(c + 3d + 3a + b)$

These are just the entries in the first four rows. We notice (without surprise, because we saw the same phenomenon in the geometrical case) that, while the four original numbers can be chosen arbitrarily, there is a relationship between the numbers in the second row: the first and third have the same sum as the second and fourth.

To see how quickly they are coming together, we need to calculate their differences. Here are the differences, arranged in the same pattern:

$a - b$	$b - c$	$c - d$	$d - a$
$\frac{1}{2}(a - c)$	$\frac{1}{2}(b - d)$	$\frac{1}{2}(c - a)$	$\frac{1}{2}(d - b)$
$\frac{1}{4}(d + a - b - c)$	$\frac{1}{4}(a + b - c - d)$	$\frac{1}{4}(b + c - d - a)$	$\frac{1}{4}(c + d - a - b)$
$\frac{1}{4}(a - c)$	$\frac{1}{4}(b - d)$	$\frac{1}{4}(c - a)$	$\frac{1}{4}(d - b)$

The difference $\frac{1}{2}(c - a)$ in the second row is matched with the difference $\frac{1}{4}(a - c)$ in the fourth row which is half as great. So once we have got past the first, arbitrary, row, the differences halve every two rows.

- It occurs to you that because of the special property of the second row, which the first row does not have, reversing the process will lead to difficulties, just as it did in the geometrical case – **turn to Frame 42.**

- It occurs to you that the numbers would decrease faster if you just took the difference in the first place, not the average. In other words, starting with the pair 10 and 12, you put down 2, whichever order they are in, rather than 11 – **turn to Frame 43.**

41

This is the result of averaging triplets of consecutive numbers, using the same starting line.

$$10 \quad 8 \quad 4 \quad 18$$

$$12 \quad 7\tfrac{1}{3} \quad 10 \quad 10\tfrac{2}{3}$$

$$10 \quad 9\tfrac{7}{9} \quad 9\tfrac{1}{3} \quad 10\tfrac{8}{9}$$

$$\cdots \quad \cdots \quad \cdots \quad \cdots$$

It does indeed seem to tend to the average faster.

- You have a reason why this should be so – **turn to Frame 45**.

- You want to investigate by calculation whether this is really true – **turn to Frame 46**.

42

If the second row consists of the four numbers marked in the second row here,

$$x$$

$$a \qquad\qquad b \qquad\qquad c \qquad\qquad d$$

and we try to make x the first number in the previous row, then the next three numbers are forced:

$$x \qquad 2b - x \qquad 2c - 2b + x \qquad 2d - 2c + 2b - x$$

$$a \qquad b \qquad c \qquad d$$

We now have to ensure that the average of x and $2d - 2c + 2b - x$ is equal to a. That is,

$$2a = x + (2d - 2c + 2b - x) \quad \text{or} \quad a - b + c - d = 0.$$

If this condition is satisfied, then any number at all can be chosen for x; but, if it is not satisfied, then no choice of x will work.

- You decide to look at the case of four geometrical points, a quadrilateral, bearing in mind that we know that the midpoints of the sides of a quadrilateral inevitably form a parallelogram – **turn to Frame 30.**

- You decide to see whether you get a similar solution working backwards from only three numbers – **turn to Frame 44.**

43

Here are two examples of the results of simply taking the absolute differences, ignoring the order of the numbers (if you calculate the differences taking order into account, so they can be positive or negative, then you get a completely different problem). In each case the numbers should be thought of as a closed chain, so that the right-hand end joins on to the left-hand end.

10	5	7	13			1	7	2	10	20
	5	2	6	3			6	5	8	10	19
2	3	4	3			13	1	3	2	9
	1	1	1	1			12	2	1	7	4
0	0	0	0			8	10	1	6	3
							2	9	5	3	5
						3	7	4	2	2
							4	3	2	0	1
						3	1	1	2	1
						...	...	...	...

- You decide to try to prove that the four-number game always eventually produces a row of zeros – **you are on your own.**

- You decide to continue investigating the five-number game, to see how it finally ends up – **you are on your own.**

- You decide to see what happens to the original averaging process if you work backwards – **turn to Frame 42.**

414 • YOU ARE A MATHEMATICIAN

44

By calculation, for the case of three numbers, to complete the first row starting with an arbitrary choice of first number, x,

$$x \qquad 2b - x \qquad 2c - 2b + x$$
$$a \qquad b \qquad c$$

we need to have $2a = x + (2c - 2b + x)$ or $x = a + b - c$. This solution is now forced, and the complete first row will now be

$$a + b - c \qquad b + c - a \qquad c + a - b$$
$$a \qquad b \qquad c$$

- The unique solution in this case, contrasted with the case of four numbers which has no solution or an infinite set of solutions, reminds you of the puzzle of finding the four missing numbers, given the sum of the numbers in pairs, as in this table:

	P	Q
R	10	8
S	7	9

The number 10 is the sum of the missing numbers P and R, and so on – **you are on your own.**

- You decide to compare working the process backwards with numbers with working it backwards geometrically – **turn to Frame 26.**

45

If we averaged four numbers at a time, all four entries would instantly become the same average, and would not change thereafter. Taking the other extreme case, if we 'averaged' one entry at a time, the original entries would never change. Filling the gaps in this pattern, it is plausible that averaging pairs will tend towards the common limit more slowly than averaging triples. Similarly, starting with, say, six numbers, averaging pairs will be slower than averaging triples, which will be slower than averaging quadruples, and so on.

- How is this conclusion proved by calculation? – **turn to Frame 46.**

- You wonder what happens if you take a quadrilateral and replace the vertices by their averages, three at a time – **turn to Frame 48.**

46

Here are the results of averaging triples of four original numbers:

a	b	c	d
$\frac{1}{3}(d + a + b)$	$\frac{1}{3}(a + b + c)$	$\frac{1}{3}(b + c + d)$	$\frac{1}{3}(c + d + a)$
$\frac{1}{9}(3a + 2b + 2c + 2d)$	$\frac{1}{9}(3b + 2c + 2d + 2a)$	$\frac{1}{9}(3c + 2d + 2a + 2b)$	$\frac{1}{9}(3d + 2a + 2b + 2c)$
$\frac{1}{27}(7a + 7b + 7d + 6c)$	$\frac{1}{27}(7a + 7b + 7c + 6d)$	$\frac{1}{27}(7b + 7c + 7d + 6a)$	$\frac{1}{27}(7c + 7d + 7a + 6b)$
. . .	. . .	. . .	. . .

The differences between the terms in the first line are $a - b$, $b - c$ and so on. In the third line they are $\frac{1}{9}(a - b)$, $\frac{1}{9}(b - c)$ and so on – one ninth as much. Similarly, the differences between the terms in the second line are $\frac{1}{3}(d - c)$ and so on. In the fourth row, the corresponding differences are $\frac{1}{27}(d - c)$ and so on, one ninth as much, so all these rows are indeed tending to the common average more quickly.

- You decide to see how the 3–2–2–2 and 7–7–7–6 number patterns continue – **you are on your own.**

- You decide to see if it always possible to calculate backwards and find the first row, given the second row – **turn to Frame 47.**

47

Whatever the values in the second row, you can calculate backwards to discover the value of the first entry, a, like this:

$$a =$$
$$\tfrac{1}{3}(d + a + b) + \tfrac{1}{3}(a + b + c) + \tfrac{1}{3}(c + d + a) - 2[\tfrac{1}{3}(b + c + d)].$$

The numbers b, c and d can be calculated similarly.

- You wonder whether a similar calculation will allow you to work backwards if you took the averages of adjacent pairs of numbers – **turn to Frame 42.**

- You wonder whether it is always possible, given a rule for constructing the second row from the first, to work backwards and reconstruct the first row from the second – **you are on your own.**

48

Figure 13.47 suggests that the averages of the vertices three at a time form a smaller copy of the original quadrilateral. Calculation confirms this: the vertices B' and C', for example, are equal to $\frac{1}{3}(A + C + D)$ and $\frac{1}{3}(A + B + D)$ respectively. The side $B'C'$ of the smaller quadrilateral is therefore, as a vector, $\frac{1}{3}(B - C)$ which is exactly one third of the original side BC, with the sign reversed.

Fig. 13.47

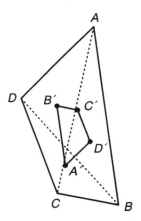

Therefore the new side is parallel to the old side and one third of its length. The same is true of other corresponding sides and so the whole new figure is the original quadrilateral rotated through 180° and reduced linearly by one third. It has the same centroid and its area is one-ninth of the original.

- You decide to investigate other properties of this figure – **you are on your own.**

- The simple relationship between the two figures suggests that reversing the process – geometrically – might be rather simple, and you decide to investigate further – **turn to Frame 26.**

Index

Please note that index references are not given to the solutions to the boxed problems, when such references would simply repeat the reference to the original box.